In Code

In Code

A Mathematical Journey

Sarah Flannery
with David Flannery

Workman Publishing • New York

 Published simultaneously in Canada by Thomas Allen & Son Limited.

Originally published in 2000 by:
Profile Books Ltd.
58a Hatton Garden
London EC1 8LX

Library of Congress Cataloging-in-Publication Data
Flannery, Sarah
In code : a mathematical journey / by Sarah Flannery with David Flannery.
p. cm.
Originally published: London : Profile Books Ltd., 2000.
Includes bibliographical references and index.
ISBN 0-7611-2384-9
1. Flannery, Sarah. 2. Coding theory. 3. Mathematicians–Ireland–Biography.
I. Flannery, David. II. Title.

QA29.F6 A3 2001
510'.92–dc21
[B] 00-069332

Workman Publishing Company
708 Broadway
New York, NY 10003-9555

www.workman.com

Printed in May 2001

10 9 8 7 6 5 4 3 2 1

Contents

Foreword

Early on the morning of Wednesday, January 6, 1999, our sixteen-year-old daughter, Sarah, set off from the village of Blarney for Dublin in the company of her science teacher, Seán Foley, and his son Vincent. Their destination was the Royal Dublin Society building in Ballsbridge, where Sarah and Vincent were to submit individual projects in the Physics, Chemistry and Mathematics category for the annual Esat Telecom Young Scientist Exhibition. The following Friday, watching the televised award ceremony at home, I was to witness with delight Vincent's winning of the prestigious Intel Award for Excellence in Science followed by the Taoiseach* (Irish Prime Minister) Bertie Ahern declaring Sarah the competition's overall winner and Ireland's Young Scientist of the Year 1999. Her award was a magnificent silver trophy, IR£1000 (about $1,400) and a one-week trip to Thessaloníki, Greece, to represent Ireland at the European Union Contest for Young Scientists in September.

Sarah's winning project, entitled "Cryptography—A New Algorithm Versus the RSA," dealt with the science of secrecy. It was given high praise by one of the competition judges during a television interview that same evening. On being asked if the project had any commercial potential, he answered, "If she plays her cards right I think she should make a lot of money . . . ." The following day one Irish newspaper headline read "JUDGES BAFFLED BY YOUNG GENIUS." Needless to say, whether or not any of these statements could withstand close scrutiny, the "lots of money/genius"

* The word itself means "leader" in Irish (Gaelic).

combination made an irresistibly attractive story for the press at large. World media attention was ensured when an article, not differing much in content from some of those that had already appeared in the Irish papers, and intended for modest inclusion somewhere in the inside pages of the London *Times,* was front-paged at the last moment. The feature was accompanied by a picture of Sarah in front of a blackboard that was covered by an array of mathematical symbols scribbled by her to please a photographer. Beneath what she regarded as this awful picture was the caption "Sarah Flannery, 16, who baffled the judges with her grasp of cryptography. They described her work as 'brilliant.'" This article sparked an explosion of worldwide interest in a young girl who a week before was leading the life of a normal sixteen-year-old.

One consequence of this article was an invitation from Profile Books, London, to both Sarah and me (as her father and mathematical mentor, who allegedly cultivated her interest in mathematics) to write a book telling a little of her life up to the winning of the award and recounting some of the events that unfolded in the months that followed. Although the book is written in the first person (Sarah), I helped to write the mathematical sections so that she could continue, as well as she could, with her schoolwork. For her mother and me, her achievement is not that she became famous for a brief period of time but that she had the interest and courage to undertake what she did. At her age I would not have been brave enough to talk with confidence to experts had I trodden but the narrowest of paths through the mathematical landscape.

David Flannery
January 2000

Preface

"Do I need to know mathematics to read this book?"

Not at all! If mathematics is not for you, we suggest that you just skim Chapters 6 and 7 to acquaint yourself with the little terminology and notation that is used sparingly in later chapters. We hope that the personal narrative in the rest of the book will interest you.

However, if you relish a challenge and can add, subtract, multiply, divide, raise a number to a power (e.g, $3^3 = 27$), and know what a square root is, then you might wish to learn, in a pleasant way, the very little extra mathematics that breathes life into the story. You will be well rewarded by acquiring an understanding of what is, fundamentally, not difficult mathematics.

We wrote Chapters 6 and 7 in hopes of keeping the book self-contained—you shouldn't have to open another book to understand this one. We have also included them for the many readers of popular scientific books who, once their interest has been sparked by the subject matter, often wish there were just a little more mathematical substance.

Anyone reading new ideas must occasionally reread a line or paragraph to recall points which may be forgotten or to clarify those which were not entirely obvious the first time around. Take the trouble to do this whenever you need to, and don't despair even if a rereading doesn't immediately do the trick. Keep going—without necessarily understanding every detail or fathoming the notation, you'll be surprised how much of the overall picture you will still absorb by simply forging ahead.

Sarah and David Flannery
January 2000

Part I
Background

1 Early Influences

There is a blackboard in our kitchen. It might be said that my mathematical journey began there. But before I tell you how I got interested in mathematics I would like to say a little about who I am. I'll just give a few basic details—by the end of this book you may know more of my character than I wished to reveal. I would also like to tell you a little about my closest relatives, who had the greatest influence on me while I was growing up.

I am the eldest of five children. I live with my parents, Elaine and David, and four younger brothers (a.k.a. "the boys"—Michael, fifteen; Brian, thirteen; David, ten; and Eamonn, seven) in an old house in the middle of a dairy farm two kilometers from the village of Blarney, County Cork. Our only neighbors, the family who own the farm, live a few hundred meters away. I attend the village's co-ed secondary school, Scoil Mhuire gan Smál. I received my primary education at the local all-girls school, and so have grown up with pretty much the same people all my life. My main interests besides math are athletics (show jumping, basketball, Gaelic football and hurling), reading, playing the piano and listening to lots of different types of music.

Both my parents are very independent-minded people. They rarely express political opinions on a party basis, but they take referenda very seriously and joke about seldom being on the winning side. They have taught us to respect the culture and beliefs of all people and to value objective truth very highly. They advise us to be ourselves and to rate moral courage above all other forms. They insist that we speak distinctly and slowly—a virtual impossibility for anybody immersed in the rich, fast-flowing Blarney brogue. Perhaps there is, after all, some truth in the legend of the

Blarney Stone, which is said to bestow the "gift of the gab" on all those who kiss it. Thousands of visitors come to perform this ritual each year, making our small village one of Ireland's most popular tourist destinations. I'd like to kiss it too, but I'm told that I talk enough as it is! I may be deluding myself, but I believe that both Mom and Dad are regarded as private but friendly people.

My mother claims that she had a wonderful childhood growing up on her family's cereal (formerly dairy) farm. She's a bit of a tomboy (she won't mind my saying so) probably because, as the eldest of nine, she had to assume the role of eldest son with all its traditional duties. This meant that from an early age she milked cows, mucked out, stacked and stored hay bales, picked stones and, best of all, was allowed to use the farm machinery to plow, rototill, harrow, mow, and harvest silage and grain. I envy her freedom to do all that, when I have to satisfy all sorts of legal and other requirements just to drive a boring old car. She's hard pressed to find use for her varied farm skills now that we have only a garden, but she still prefers to be outdoors. She left home at eighteen for university, where she studied microbiology, a subject about which she is most passionate. She lectures part-time at the same institute as Dad—the Cork Institute of Technology (CIT), eight kilometers away. She is very cool and has a calming influence on us all. She loves writing limericks, mostly about my father's idiosyncrasies—he appears not to mind being the butt of these well-crafted lines. At night she likes to take out her guitar and sing. She has a fine voice, and whenever I see her and my brother Michael playing guitar and singing together, I make sure Dad knows how much I hate him for the fact that I have inherited his crowlike voice. Mom and Michael like to harmonize, and lately they've been doing songs from Bob Dylan's *Desire* album.

My father's passion is mathematics, so he spends a lot of time in the sitting room "studying." His enthusiasm must rub off because Mom has been cornered many times by students who want to tell her what a great lecturer they think Dad is, and how "he makes math come alive." He is not into sports, but he played poker for years to bring in extra income. He loved chess as a student and has taught all of us how to play, even though we much prefer outdoor sports. He has a few devoted friends, mostly intellectual types with liberal views, but hates socializing, which is a pity as my mother is very fond of people and civilized conversation. He can, however, spend hours talking in a pub, particularly with his brothers, who relish verbal sparring on any issue. I've always enjoyed listening to these conversations in the grown-up atmosphere.

My brother Michael lives for the day when he can deliver a knockout piece of repartee to Uncle Paul, a barrister much given to biting sarcasm but who admits that it is only a matter of time before we younger ones gain the upper hand. This good-humored uncle once said to me in frustration, "Would you ever go away and develop a few complexes?!" Dad thought this was hilarious, and a high compliment to me.

My mother's father is the most independent-minded man I've ever encountered and a legend among those who know him as a "genius with machinery." Once, for example, when he needed a more powerful tractor but couldn't afford it, he made one. He bought a truck engine and completely modified his tractor to make the new engine fit. My mother says that she remembers people coming from miles around to watch this undertaking and to try out the finished product on nearby Barleymount Hill, an incline so steep it forces all vehicles to downshift but which made no impression on the new tractor as it soared up in high gear. He is largely a self-educated man who came, after much thought, to be an avowed atheist at a time when it was uncomfortable and unwise to be a nonbeliever in a small, predominantly Catholic farming community. My father, an agnostic, teases him that he's as dogmatic about his own beliefs as any of the old-style priests he so criticizes, but my grandfather replies that agnostics are wimps who are afraid to face the obvious. (My father and my grandfather get along well because they value "things of the mind" but never take themselves too seriously.) My grandfather has a passion for astronomy and a fascination with everything that is either "the biggest" or "the longest" or "the strongest," but has no interest in the "smallest," and is totally incapable of making small talk. He can bring any conversation, no matter how varied its participants, around to one of his pet topics in no time at all and insists on getting everyone's opinions on the most arcane matters whether they are interested or not.

His wife, my maternal grandmother, must be a saint, having lived with such a man all her adult life while trying to raise their nine children. She loves traditional Irish music and set dancing,* and is the stereotypical welcoming and generous farmer's wife who never lets anyone out her door without first having tea and her famous soda bread. She has made many a witty remark to put my grandfather in his place. On one occasion, after he had told a visitor how much he would like to live in a Mediterranean climate and enthused about the siesta custom ("I'd love that—a sleep in the middle of the day!") my grandmother peeped around the door and chided him with, "You wouldn't be up in time!" (He stays up so late on winter nights watching the stars or reading about them that he is not always an early riser.)

My father's father started as a railway clerk at seventeen, paying more in rent than he was earning, and retired forty-three years later as a chief executive of the company. He and his father before him loved mathematics and were noted for their natural ability at it. My great-grandfather was always being called upon to "step out" land for the purpose of ascertaining its area. My grandfather says that his father's brother, known as Yellow Jack because of his jaundiced complexion, was regarded as an even better mathematician. These men, with no more than a primary education, were not mathematicians in the modern sense but they were gifted calculators and knew some geometry. My grandfather scored 99, 97 and 100 per cent, respectively, in arithmetic, algebra and geometry for his Intermediate Certificate at a time when these exams were much harder than they are now (or so he tells us). I remember him telling me of how he saw a mason using strings of lengths 3, 4 and 5 to form a right-angled triangle, which he said was a practical use of the converse of the Pythagorean theorem.

*An elaborate dance consisting of five separate parts, danced by four couples.

He loves puzzles of all types. He always did the *Irish Times* Simplex crossword, but when he retired he set about learning how to do the cryptic version—the Crosaire. He mastered this art in a few weeks after many exasperating efforts, tenaciously examining the published solutions on the following day. Now he instructs my mother in how to decipher their mysteries, which my father, who refuses even to try them, describes as the product of a warped mind.

Married for fifty years, he and Grandma had eight children of whom my father is the eldest. I know that most women of her generation stayed at home to rear the children, but Grandma sacrificed more than her own career to support Granddad in his. Due to his promotions in the company, she had to pack the bags, uproot her children, say goodbye to neighbors and friends and move to a different town, setting up house no fewer than ten times. She is noted for a deep intuition about people and is rarely wrong in her assessments. She is an extremely efficient housekeeper, often preparing four-course dinners for as many as fourteen without apparent effort. Everybody else gets to do the washing-up. I often marvel at her boundless energy and zest for life. Since Granddad retired, the two of them seem to be on holiday as often as they are at home. They're avid cardplayers and formidable bridge partners, playing several nights a week. Dad jokes that they unwittingly steered him towards single-person games such as chess and poker where you have nobody to blame but yourself if you don't win. None of us knows how to play bridge, though my grandparents are always threatening to fix that, despite whatever reservations any of us might have. Now that you know what stock I come from, you won't be surprised to learn that I, too, have a liking for games, puzzles and things mathematical. In the next chapter, I'll tell you about a method my parents, in particular my father, used when I was younger to arouse intellectual curiosity in me and my brothers, and to help us think.

2 Early Challenges

I didn't get any special tutoring in math while I was growing up. I was left to my own devices. I could, and did, ask for help with homework now and then, and I still do. My brothers are free to do the same, though I suspect that some of them would rather die than ask for help. The reason for this is simple: In the Flannery house, you often get more help than you ask for. You run the risk of a little homily on how nice a poem is, or "Isn't it very clever how that is done," when all you want is the question answered quickly so you can get your homework out of the way and be free!

Strictly speaking, it is not true to say that I or my brothers don't get any help with math. We're not forced to take extra classes, or endure grueling sessions at the kitchen table, but almost without our knowing we've been getting help since we were very young—out-of-the-ordinary help of a subtle and playful kind which I think has made us self-confident in problem-solving. Ever since I can remember, my father has given us little problems and puzzles. I have often heard, and still hear, "Dad, give us a puzzle." These puzzles challenged us and encouraged our curiosity, and many of them made math interesting and tangible. More fundamentally, they taught us how to reason and think for ourselves. This is how puzzles have been far more beneficial to me than years of learning formulae and "proofs."

One of the first puzzles I remember hearing is one I'm sure every child has heard in one form or another. A farmer, on his way home with a fox, a goose and a sheaf of oats, comes to a narrow wooden bridge that can support only the weight of the farmer and *one* of his charges. If the farmer leaves the fox alone with the goose while he carries the sheaf of oats across the bridge, the fox will eat the goose. If he brings the fox across and leaves the goose alone with the sheaf of oats, the goose will eat the oats. So how does the farmer get himself, the fox, the goose and the sheaf of oats safely to the other side? I won't tell you how it's done, but just in case you mightn't know, let me say that a fox has no interest whatsoever in oats.

Here are a dozen or so model puzzles to show some of the many lessons I learned and how my mind was shaped by them. The puzzles are not too difficult, and I give the answers to illustrate the points I want to make. But try your best to solve them before you skip down to the solutions.

Many puzzles are simply numerical in nature, as is the one that follows. My dad gave it to me when I was quite young, probably about five years old. It crops up in the film *Die Hard with a Vengeance*. The inept attempts of the characters played by Bruce Willis and Samuel L. Jackson at a solution had one of my brothers nearly on top of the TV trying to tell them the answer.

The Two Jars Puzzle: *Given a five-liter jar and a three-liter jar and an unlimited supply of water, how do you measure out four liters exactly?*

Bear in mind that we, my brothers and I, are given puzzles under less harrowing conditions than those in the film, and solving them is not a matter of life or death for us.* But if you listened in on some of our conversations, you could be forgiven for thinking otherwise. You'd often hear a bad-tempered "Be quiet, will you, I'm thinking!" hurled by a younger sibling at an older one who is acting all superior and saying things like, "That's simple" or "That's too hard for you, dumbo." Needless to say, the arrogant older brother has heard the puzzle before and knows (or is it remembers?) the answer. But that's OK—he'll be given a real brain twister to put him in his place.

*On the off chance that you haven't seen the movie, the two heroes must measure out exactly four liters in under five minutes to stop a bomb from going off and blowing them to pieces. Stressful to say the least, and definitely a matter of life or death!

Dad likes to make the most of a simple puzzle by extending it a little whenever possible. For example, let's say you've solved the two-jar problem and you're hoping to be rewarded with another puzzle of a different type; you are more likely to hear, "Now, see if you can measure out exactly one liter." Easy enough, but once you do it (and presuming you have not become bored) he will then suggest that you try the same problem with nine- and four-liter jars to get all the possible measures between one and thirteen liters. Of course this is a little more taxing, but it gets you beyond trial and error to actually thinking. When you eventually succeed in doing it, you are then asked, "How might you do this most efficiently?" At this Mom says, "Mathematicians can never leave well enough alone—what if this, what if that? They take the good out of it. You never get to the end of anything." She's right, but the process can be great fun. This is perhaps what makes the puzzles so worthwhile: some have the germs of greater things in them. When presented to enthusiastic children hungry for challenges, these puzzles can teach a lot about logical thinking, while providing amusement.

So when you tire of all this careful water-pouring and think that surely now you've earned a fresh puzzle you might hear, "I'll give you a new puzzle, but first tell me how to measure exactly four liters with only a six-liter jar and a three-liter jar." "Ahh, Dad!" What's worse is that after a while you realize it can't be done. This variation on a theme lets you learn the very valuable lesson that sometimes problems may not have solutions of the type you seek (and that you cannot trust even your own father to give you solvable puzzles). Sometimes when I am really stuck I'll take a chance with, "I'll bet this is one of your stupid has-no-solution puzzles," but this still doesn't always put an end to the agony of not being able to find an answer. I must say, though, that it is a pretty good way to fish for clues when things get just a little too frustrating and you're losing your will to keep searching. But before you become desperate, here's the solution to the two-jars puzzle:

Fill up the five-liter jar. Then use it to fill the three-liter jar so that two liters remain in the five-liter jar. Empty the three-liter jar and pour the two liters from the five-liter jar into the three-liter jar. Fill up the five-liter jar again and use it to top up the three-liter jar with the extra liter it needs. Four liters now remain in the five-liter jar.

Alternatively, fill up the three-liter jar and pour its contents into the five-liter jar. Fill up the three-liter jar again and use it to top up the five-liter jar with the extra two liters it needs. Now empty the five-liter jar, and pour the liter that is in the three-liter jar into the empty five-liter jar. Fill up the three-liter jar again and pour its contents into the five-liter jar. The five-liter jar now contains exactly four liters.

My youngest brother, Eamonn, says, "Start filling and pouring and just keep going." He's right. You will eventually end up with what you were asked to get, provided it's possible. With a nine-liter jar and a six-liter jar you cannot measure five liters exactly. On the other hand, with a nine-liter jar and a seven-liter jar it is possible to measure out exactly every measure from one to nine liters.

A variation on this Two Jars Puzzle is the **Egg Timer Puzzle:** *How would you boil a four-minute egg with two egg timers that measure three and five minutes, respectively, and what is the shortest amount of time required?* You'll have to answer this one for yourself.

Sometimes Dad comes up with really good puzzles that drive us up a wall trying to figure out what could possibly be wrong with our seemingly logical answers while he just sits chuckling at the head of the table. Watch out for puzzles where the first answer that comes to mind just seems too easy. The following, which is also in the purely numerical vein, should convince you that sometimes $3 - 2 \neq 1$. (The symbol $\neq$ means "is not equal to.")

The Rabbit Puzzle: *A rabbit falls into a dry well, thirty meters deep. Since being at the bottom of a well was not her original plan, she decides to climb out. When she attempts to do so she finds that after going up three meters (and this is the sad part) she slips back two. Frustrated, she stops where she is for that day and resumes her efforts the following morning—with the same result. How many days does it take her to get out of the well?*

Don't jump to conclusions. Once you see and solve (or just hear the solution to) a problem of this type, you learn to be careful and watch out for little wrinkles:

> Because the rabbit goes up three meters every day and then slides back down two, she gains a meter every day. However, it would be an error to conclude that she escapes from the well after thirty days. The manner in which she gains the meter per day is all-important. After twenty-seven days she is three meters from the top. On the twenty-eighth day she reaches the top and runs off. It takes the rabbit only twenty-eight days to get out of the well.

See if you have learned the lesson of being cautious by trying this next puzzle.

The 100-Meter-Dash Puzzle: *If A, B and C run a 100-meter dash, each running at a uniform speed throughout, and if A beats B by 10 meters and B beats C by 10 meters, by how much does A beat C?*

Now cover up the solution below while you try to figure it out yourself—you can do it!

> When *A* crosses the 100-meter line to win, *B* is at the 90-meter mark because he covers 9 meters for every 10

meters that A does. Similarly, since B beats C by 10 meters, C covers only 9 meters for every 10 meters that B does, so when B has covered 90 meters, C has covered only nine-tenths of 90 meters, which is 81 meters. So when A crosses the 100-meter line to win, C is at the 81-meter mark. Thus A beats C by 19 meters.

When you finally crack a problem, whether by yourself or with some help, you experience a great feeling of self-satisfaction and pride. Recently Dad asked my ten-year-old brother, David, "What is the smallest number with *exactly* seven factors?" This kept him quiet for a while. In fact, the puzzle was a little too hard for him (or at least it did not pique his curiosity) and he appeared not to be thinking about it because a day or two later I heard him ask for a different puzzle, only to be told, "But you didn't solve the last one I gave you." When David made some evasive excuse, Dad—contrary to his usual practice of never giving the least hint—took him to the kitchen blackboard and got him to do some calculations. Then he gave him a little nudge in the right direction (factors coming in pairs). Then I saw David's eyes light up: "Oh, yeah!" In no time at all he scribbled the answer on the board and turned around, smiling like a Cheshire cat. "That was cool," he said, and skipped off to his next pursuit.

Now here is my favorite puzzle. It requires only simple arithmetic skills, but the thought processes of a detective. I especially like the look that develops on people's faces as I start to tell it—a look that by the end shows that they certainly think I'm cracked. See what you think when you first read it.

The Insurance Man Puzzle: *An insurance salesman knocks on the door of a home in a housing development. When a lady answers he asks, "How many children do you have?" She replies, "Three." When he asks, "What are their ages?" she decides that he is too cheeky and refuses to tell him. After he apologizes for his apparent rudeness he asks for a hint about the children's ages. She says, "If you multiply their three ages you get 36." (Their ages are exact numbers.) He thinks for a while* *and then asks for another hint. When she says, "The sum of their ages is the number on the house next door," he immediately jumps over the fence to determine this number. This done, he returns to the lady and asks for one last hint. "All right," she says, "the eldest plays the piano!" He then knows their ages. Do you?*

It really isn't a very difficult problem, but I like it because it is an exercise in methodical, or vertical, thinking. Study the facts as they are presented, think about them logically, and do not be distracted by talk of piano players. I discuss the solution a little further on.

Meanwhile, here is a completely different problem requiring no numerical skills but which tests observation and might appeal to those of a literary bent:

The Text Puzzle: *Study this paragraph and all things in it. What is vitally wrong with it? Actually, nothing in it is wrong, but you must admit that it is most unusual. Don't just zip through it quickly, but study it scrupulously. With luck you should spot what is so particular about it and all words found in it. Can you say what it is? Tax your brains and try again. Don't miss a word or a symbol. It isn't all that difficult.*

Hint: cryptanalysts (the guys who try to crack cryptosystems —when they work for us we think of them as very clever fellows, but when they work for the enemy we think of them as low-down sneaks) would spot the answer immediately.

Do not read on unless you have tried very hard to solve these puzzles.

Let's begin with the insurance man puzzle. You might reason that one child is aged 2, and another 3, so that the remaining child is 6, since $2 \times 3 \times 6 = 36$. Thus you have found one possible "triple" {2,3,6}. You might then discover that another solution is {2,2,9}. At this point you'll probably adopt a more systematic approach to finding the possible ages. You could start by asking, "What is the youngest possible age a child can be?" Since we are dealing with exact numbers the answer to this question is 1. Given this age for one of the children, what could the next child's age be? It could also be 1, leaving the other child at the ripe old age of 36. Unusual, extremely unlikely, but a possibility. Continuing in this way, you will come to realize that after the first hint the insurance man knows that the correct triple of ages is one of *eight* possibilities since

$$\begin{aligned}
1 \times 1 \times 36 &= 36\\
1 \times 2 \times 18 &= 36\\
1 \times 3 \times 12 &= 36\\
1 \times 4 \times 9 &= 36\\
1 \times 6 \times 6 &= 36\\
2 \times 2 \times 9 &= 36\\
2 \times 3 \times 6 &= 36\\
3 \times 3 \times 4 &= 36
\end{aligned}$$

It is now clear why he had to ask for another hint. He simply does not have enough information at this stage to choose the correct triple from this set of eight. However, we know that he was told that the sum of their ages is the number on the house next door, and we complain bitterly that he was able to go look at this number but we don't get to see it. But the great thing about this puzzle is that we do see it—not physically, but with certainty in our minds. To find out why, let's ask what numbers he could have seen. Here they are in the column on the right:

$$\begin{aligned}
1 + 1 + 36 &= 38\\
1 + 2 + 18 &= 21\\
1 + 3 + 12 &= 16\\
1 + 4 + 9 &= 14\\
1 + 6 + 6 &= 13\\
2 + 2 + 9 &= 13\\
2 + 3 + 6 &= 11\\
3 + 3 + 4 &= 10
\end{aligned}$$

Now what do you notice? Examine the possible sums carefully. As the textbooks often say, "the alert reader will note" that the sums are all distinct with the exception of the number 13. There are two triples whose sums are 13. Now what was the number on the house next door? It was 13! Had he seen the number 21 he would have known then and there that the ages of the children are 1, 2 and 18 and he would not have needed an extra clue. So he must have seen the number 13. Then the possibilities for the ages are {1,6,6} or {2,2,9}. We now see the relevance of that final enough-to-put-anyone-off-clue, "The eldest plays the piano," and note that the key word is "eldest." The family of three children whose ages are 1, 6 and 6, respectively, does not have an eldest. Therefore the children are aged 2, 2 and 9. Now, isn't that a nice piece of detective work?

So are you a potential cryptanalyst? The answer to the word puzzle is this:

> Not one single word in the paragraph contains the most frequently occurring letter of the English alphabet, *e*. Furthermore, every other letter of the alphabet occurs at least once.

Try this "cryptographic" puzzle:

The Russian Postal System Puzzle: *I heard this interesting puzzle posed by a Russian mathematician at the end of a popular talk on secrecy he gave one night at our local university. He told us that at one time the Russian postal system was notoriously corrupt. Any letter, package or box which was open or easily openable would be opened in the sorting office, and anything inside would be removed whether or not it had any value. However, since the pickings were so rich the sorters never bothered to open anything that was locked, even if they suspected it contained valuables.*

Now, Boris in Moscow had bought a beautiful gem for his girlfriend, Natasha, who lived in St. Petersburg, and he wanted to get it to her as quickly as possible. Neither he nor Natasha could travel to the other's city, so what was he to do? He had a strongbox with a hasp to which a number of padlocks could be attached. If he bought a padlock and key he could put the gem in the box, lock the padlock and send the box through the postal system knowing that it would not be pried open and that it would be delivered to his beloved. But what good would that do? Natasha would not have a key to open the padlock. Boris couldn't send the key separately by letter as it would be opened and the key removed. However, Boris phoned Natasha and between them they hatched a clever scheme by which they could get the precious jewel from Moscow to St. Petersburg in safety despite the corrupt postal system. How did they do this?

Here's the solution, but resist the impulse to read it without even trying to figure it out.

> Boris in Moscow rings Natasha in St. Petersburg and tells her that he will place the gem in the strongbox, lock it with a padlock and post it to her. Because it is locked it will arrive safely in St. Petersburg. She is then to lock the strongbox a second time with a padlock of her own and post it back to him in Moscow. He will then remove his lock and send the strongbox, with the gem still safely locked inside, back to her again. This time when it arrives safely in St. Petersburg she can open it. Neat!

The great thing about puzzle-solving is that it is not always the "professional" who first finds the solution to a puzzle, or, if finding a solution, discovers the most elegant and imaginative path to its unveiling. Solving a puzzle is undertaking a journey and reaching a destination. Lacking the "vertical" training of the academic, the amateur by necessity must be more creative and is often rewarded by finding the shortest and most beautiful route.

To illustrate this point I present two problems: one I heard in relation to the phenomenal calculating ability of the Hungarian mathematician John von Neumann; the other is mentioned in Arthur Koestler's book *The Act of Creation*. Here is the first:

The Two Trains Puzzle: *Two trains are on the same railroad track 100 km apart and heading towards each other, each at a speed of 50* *km/h. A fly, initially on the front of one train, flies at 75 km/h towards the other oncoming train. On reaching it, the fly turns around (instantaneously) and flies back towards the first train. When it reaches the front of this train it turns around (again instantaneously) and flies back in the other direction. How many kilometers will it have traveled in this hectic zigzag manner before it meets its inevitable fate?*

Von Neumann is reputed to have answered "75 kilometers" immediately. When asked how he did it, he replied, "I summed the infinite series, of course!" The questioner, instead of being surprised that the great man had not found the answer the "neat way," was stunned that Von Neumann could see instantly that the fly traveled the ever shorter paths measuring 60, 12, 2.4, 0.48, 0.096, . . . , and that he had added this infinite sequence of numbers to get a total distance traveled of 75 km. What chance have the rest of us got? We may console ourselves by thinking that we would have found the smart solution: since the trains will collide in an hour, the fly traveling at 75 km/h will cover exactly 75 km. Now isn't that clever!

I'll let Arthur Koestler tell the next one in his own words.

The Buddhist Monk Puzzle: *One morning, exactly at sunrise, a Buddhist monk leaves his temple and begins to climb a tall mountain. The narrow path, no more than a foot or two wide, spiraled around the mountain to a glittering temple at the summit. The monk ascended the path at varying rates of speed, stopping many times along the way to rest and eat the dried fruit he carried with him. He reached the temple shortly before sunset. After several days of fasting he began his journey back along the same path, starting at sunrise and again walking at variable speeds with many pauses along the way, finally arriving at the lower temple just before sunset. Prove that there is a spot along the path that the monk will occupy on both trips at precisely the same time of day.*

This is a completely different puzzle from those you've seen so far. There are no numerical facts, such as the monk's speed, with which to begin calculating. In fact, we have no specific details of the monk's ascent or descent other than that, in both cases, he leaves at sunrise and arrives just before sunset. The time span of both journeys is the same. A trained mathematician might eventually be led to solve this problem graphically. Koestler described the following solution, by a young woman with no scientific training, as brilliant.

> I tried this and that, until I got fed up with the whole thing, but the image of the monk in his saffron robe walking up the hill kept persisting in my mind. Then a moment came when, superimposed on this image, I saw another, more transparent one, of the monk walking *down* the hill and I realized in a flash that the two figures *must* meet at some point some time—regardless at what speed they walk and how often each of them stops. Then I reasoned out what I already knew: whether the monk descends two or three days later comes to the same; so I was quite justified in letting him descend on the same day, in duplicate so to speak.

The next puzzle is similar to the one just discussed and might leave you wondering, where do I start? While it has its appeal, I must admit that I often find this type of puzzle very frustrating because you either see how to do it or you don't. I prefer those numerical puzzles that allow you to get a hold straightaway. As soon as you begin to investigate, lines of inquiry suggest themselves. Pursue them logically and methodically and the puzzle will eventually unravel and reveal its solution.

I know many people revel in the *either you see it or you don't* form of brainteaser because they confidently believe that, with some thought, inspiration will strike. Whether it does or not, I am always impressed when the reasoning used to find the answer is explained.

The Twenty People at a Party Puzzle: *Suppose there are twenty people in a room. If Alice and Bob are any two of them, and Alice knows Bob, then you may assume that Bob knows Alice. Fur-*

thermore, if Alice does not know Bob then, likewise, Bob does not know Alice. Now, any individual among these twenty may know nobody else, or some but not all of the others, or know everybody in the room. However, what might strike you as amazing on first thought is the fact that all twenty cannot each know a different number of people in the room. Put another way, there are at least two in the room who know exactly the same number of people. Can you reason out why this must be so?

Here's why, but once again I urge you not to read it before you puzzle it out on your own.

> Imagine each of the people at the party wearing a T-shirt with a number written on the front of it. This number tells how many people at the party this person knows. So the guy with the number 4 on his shirt knows exactly four people (outside of knowing himself, which doesn't count). Someone else with the number 11 knows exactly eleven people at the party. Since the largest number of people any one person can know is nineteen (all the others), the largest possible number that can be seen on any T-shirt is 19. Similarly, since the smallest number of people any one person can know is zero (none of the others), the smallest possible number that can be seen on a T-shirt is 0.
>
> However, if there is someone in the room with the number 19 on his shirt (meaning that this person knows everybody else), then there cannot be a person in the room with the number 0 on his shirt. And if there is someone in the room with the number 0 on his shirt (meaning that this person knows nobody else) then there cannot be a person in the room with the number 19 on his shirt. This is the key observation. In either case, there are only nineteen different numbers that can appear on T-shirts, either those ranging from 1 to 19 or those ranging from 0 to 18. Since there are twenty people in the

room, two people must have the same number on their shirts and so know the same number of people.

This next puzzle pits intuition against logic.

The Rope Around the Earth Puzzle: *Imagine a rope tied around the Earth's equator like a ring on a person's finger. Now imagine lifting off this very long rope (don't ask me how), cutting it somewhere so as to stitch into it exactly one meter of extra rope. Then imagine placing this longer rope back around the Earth at the equator. Since it is longer than the original rope by just one meter, a gap between the rope and the Earth's surface will form all the way around. (Don't trouble yourself about the rope "falling down.") How large is the gap that is formed?*

Give an intuitive answer to this puzzle and then think carefully about it for a while. Many people find the answer so astounding that even when it is explained they are apt to say something like, "I see it but I still don't believe it."

The answer is very close to a staggering 16 cm, a little wider than this page, and obtaining it requires nothing more than knowing that the circumference of a circle of radius R is $2\pi R$. If we let R be the radius of the Earth and g be the gap size, both measured in centimeters, then (the symbol $\Rightarrow$ stands for "implies that"; the symbol $\approx$ stands for "is approximately")

$$\begin{aligned} 2\pi(R+g) &= 2\pi R + 100 \\ \Rightarrow 2\pi R + 2\pi g &= 2\pi R + 100 \\ \Rightarrow 2\pi g &= 100 \\ \Rightarrow g &= \frac{100}{2\pi} \approx 16 \end{aligned}$$

Many find this answer incredible and wonder how adding just an extra meter can have such an effect. But as this solution reveals, the answer is independent of the length of the original rope and is one you are unlikely to forget. If you tied a rope around a basketball instead of the Earth, and added an extra meter as before, the resulting gap would still be approximately 16 cm.

Magic Squares

I now present my final puzzle, the one from which I learned the most. Later, when I started to give talks, I found it to be the best illustrator of some of the many lessons one can learn from puzzle-solving.

Place the numbers 1 to 9 into the nine cells of a three-by-three square:

–	–	–
–	–	–
–	–	–

so that the sum of the entries along

- *each of the three rows*
- *each of the three columns*
- *each of the two diagonals*

is 15.

"Can it be done?" I hear you ask.

"Amazingly, yes," I answer with confidence.

A square in which each of the eight possible sums is 15 is known as a three-by-three magic square, and the sum 15 is said to be the magic sum. I'll tell you later how to figure out why this magic sum is 15 and not some other number.

The square

5	6	4
9	7	8
1	2	3

is not a magic square because it does not meet all eight "magic conditions." Although its three columns sum to 15, one row sums to 15 and one diagonal sums to 15, there are two rows and one diagonal whose sums are *not* 15.

As a start to solving this problem, simply use trial and error. Do not give up if you do not find a solution quickly, but keep trying—remember, it can be done.

After many attempts to place the nine numbers in the square you might begin to realize that the central cell holds the key to solving this puzzle. This cell is *unique* in that it is the only cell that touches all the other cells:

–	–	–
–		–
–	–	–

Consequently, whatever number is placed in this cell gets added to each of the remaining numbers along some row, column or diagonal. This is an important observation. Let us see which numbers can be placed in this cell. Can the number 6 be placed in the central cell, as shown?

–	–	–
–	6	–
–	–	–

No! Why not? Think carefully before reading any further.

If you are still at a loss, look at the diagram and ask yourself, "If 6 is in the central cell, where can I place the number 9?"

Nowhere! The number 9 must be placed in one of the surrounding cells, but when added to the 6 in the center, a sum of 15 is obtained with just two numbers. This observation is a big breakthrough because, for the very same reason, the central cell cannot contain the number 7, 8 or 9 as two numbers alone would then sum to 15 or more, along one of the rows, columns or diagonals. The trouble is simply that each of the numbers 6, 7, 8 and 9 is too big to go in the center. Now, isn't that a nice piece of reasoning, blending simple arithmetic with the geometry of the cells?

Can any of the numbers 1, 2, 3 and 4 be placed in the central cell? Maybe these numbers are too small.

Let us be specific. Can we place 4 in the central cell? No, for then where can the 1 go? Nowhere, because 1 + 4 = 5 leaves us looking for a 10 to make a sum of 15, and "there ain't no 10." This is great. Better still, the situation is the same with 3, 2 or 1 in the center. Thus we have discovered that neither 6, 7, 8, 9 nor 4, 3, 2, 1 can go in the central cell.

Using a little logic, we now know that our only chance of success is to place the middle number 5 in the central cell:

–	–	–
–	5	–
–	–	–

Doing this does not guarantee success, but it is necessary. It remains to be seen if it will be sufficient. Even at this stage we have learned a valuable lesson: "A little thinking can save a lot of computing."

Knowing that the number 5 must go in the central cell is a huge step forward. Rather than resume the trial-and-error approach, try to make further progress by doing some more thinking. Think of it as smart detective work.

Maybe there are restrictions on where some of the other numbers may be placed. The number 9, being the biggest, might be worth further investigation. Where can the number 9 be placed? In a corner cell?

Let us see if we can place the number 9 in a corner cell:

9	–	–
–	5	–
–	–	*

If so, the cell marked with an * must contain the number 1. Why? (See Appendix B, page 297, if you don't know why.) Thus we have the following diagram:

9	–	–
–	5	x
–	x	1

Now here comes a killer observation: the numbers 6, 7 and 8 are too big to share a row or column with 9 so they must occupy the two cells marked with x. But "three into two won't go": we cannot place three numbers in two cells. We have made another major step forward. The number 9 *cannot* be placed in a corner cell.

So, using simple reasoning and the minimum of computation, we now know that the number

5 must go into the central cell
9 must go into one of the central side cells

Try

A	–	–
9	5	1
B	–	–

where, in the first column, we have labeled the companion entries to 9 as A and B for further scrutiny. Resist the temptation to stop reasoning and finish the puzzle by testing a few numbers. Do a little more detective work and you'll see why it's so worthwhile.

The numbers *A* and *B* in the first column cannot be any of 6, 7 or 8. (Once again, see Appendix B, page 297, if you want to know why.) Hence they must be 2, 3 or 4. But *A* cannot be 3 because then *B* would have to be 3 also, which is impossible since 3 can be used only once. Thus *A* must be 4 or 2 (in which case *B* is either 2 or 4, respectively).

Victory is in sight. Let's choose *A* to be 4. We are now at this stage:

4	–	–
9	5	1
2	–	–

Now, where all the remaining numbers go is forced, so we may say that the square "completes itself":

4	3	8
9	5	1
2	7	6

Pure magic!

After I first solved this problem by trial and error, Dad took me through the puzzle slowly just as I have done above. He wanted to show me how sometimes "things can be reasoned out." It was a wonderful lesson and convinced me of the importance of standing back from a problem to think about it rather than rushing into it. Sometimes I still just like to dive in because it's great fun fiddling around with numbers, but if you really want to make inroads on a particular problem it is better to stop and reflect. It was fantastic to see what pure thought alone could achieve. I immediately wanted more problems of this type, but Dad said that there was still more to be learned from this magic square puzzle. In fact, the best was yet to come: with a little more thinking we could say exactly how many three-by-three magic squares there are in all. This unexpected bonus would come as a result of thinking the problem through from beginning to end, without running off at some stage to finish it by plugging in numbers until we found a combination that worked. Had we done this we'd have missed out on seeing deeper into the problem and proving, almost without realizing it, the "little theorem" which says that there are exactly eight three-by-three magic squares which use the numbers 1 through 9.

Why eight? I asked myself. Well, let's see. For starters, there is no choice in where the 5 must go: it must always be placed in the central cell. However, there are exactly four choices of cell for the number 9. When this is chosen the position of the number 1 is automatically determined, as are the two possible cell choices for the number 4. When a cell is chosen for the number 4 there are no further cell choices for the remaining numbers. Since any of the first four choices for the 9 can be combined with any of the two choices for the 4, there are in all $4 \times 2 = 8$ possible three-by-three magic squares. That's all there is to it, but what a thrill to know why!

Here are the eight solutions:

1.

8	1	6
3	5	7
4	9	2

2.

6	7	2
1	5	9
8	3	4

3.

2	9	4
7	5	3
6	1	8

4.

4	3	8
9	5	1
2	7	6

5.

4	9	2
3	5	7
8	1	6

6.

6	1	8
7	5	3
2	9	4

7.

2	7	6
9	5	1
4	3	8

8.

8	3	4
1	5	9
6	7	2

These eight solutions are related to one another. Suppose the original solution we found was:

8	1	6
3	5	7
4	9	2

Then the first four solutions listed are obtained from this solution by leaving it as it is or rotating the square counterclockwise through 90°, 180° and 270°, respectively. (Of course when you do this you have to "stand the numbers right-side up" again!)

Thus

1.

8	1	6
3	5	7
4	9	2

2.

6	7	2
1	5	9
8	3	4

3.

2	9	4
7	5	3
6	1	8

4.

4	3	8
9	5	1
2	7	6

are the solutions obtained by rotating the original solution. The second set of four solutions,

5.

4	9	2
3	5	7
8	1	6

6.

6	1	8
7	5	3
2	9	4

7.

2	7	6
9	5	1
4	3	8

8.

8	3	4
1	5	9
6	7	2

is obtained from the original solution by reflections in the middle row, middle column and in the two diagonals. In the first of these squares (#5), the middle row is the same as the middle row of square #1, while in the second (#6) the middle column is the same as that of the original square (#1). In the third (#7) the diagonal running from SW to NE is the same as the corresponding diagonal in the original square, while in the fourth (#8) the diagonal running from NW to SE is the same as that in the original square. Notice that the number 5 never leaves the central square under any of the four rotations or reflections.

From a geometrical viewpoint the eight solutions displayed above can be considered as one because once any one of them is found, the other seven are obtained from it using the symmetries of a square. This is why mathematicians say that "up to symmetry" there is only one three-by-three magic square. I like the nice way arithmetic and geometry blend in this puzzle.

To finish off, let us answer the question,

Why is 15 the magic sum for a three-by-three magic square?

Suppose the magic sum is M. Then the entries across the top row add up to M, as do the entries across the middle row and the bottom row. When you total the sums for the three rows you get $3M$. However, when you total the three rows together, you are also adding the first nine natural numbers in some order:

$$3M = 1 + 2 + 3 + 4 + 5 + 6 + 7 + 8 + 9$$
$$\rightarrow 3M = 45$$
$$\Rightarrow M = 15$$

Voilà!

You can appreciate fully what has been accomplished when I tell you that there are

$$9 \times 8 \times 7 \times 6 \times 5 \times 4 \times 3 \times 2 \times 1 = 362{,}880$$

different ways to place the numbers 1 to 9 into the nine cells of a three-by-three square. Because each square gives rise to eight configurations, one-eighth of this number, 45,360, is the number of genuinely different squares. Only one of these is a magic square. It is intriguing that there is one and only one magic square. After all, there was no reason at the outset to believe that there had to be even one. Or was there?

This problem is a fine example of how, with reasoning, a multitude of possibilities can be narrowed down to just a few candidates.

You could solve this problem by programming a computer to generate all 362,880 possible squares, test the eight row, column and diagonal sums of each square and select only those for which all eight sums are 15. Although the enormous computer calculation will give the eight solutions, it can never explain why only eight are found. To answer this, you, the human being, have to think about the problem.

Puzzles, like humor, have a universal appeal and know of no boundaries—cultural, educational or otherwise. People of all ages and levels of education are attracted to the puzzle as they are to the joke. In a sense, there is an affinity between the two in that a vital ingredient of both is the element of surprise. No problem is worthy of the name "puzzle" if its solution is obvious, just as the joke whose punch line is easily anticipated is soon forgotten. The true puzzle should be accessible to all; its solution should require no special knowledge other than, at times, the rudiments of arithmetic and algebra. It is perhaps the unconscious feeling that we all start out equal that gives puzzles their charm.

Later, when the cryptographic project Dad mentioned in the foreword became the subject of media attention, someone asked me how I had the confidence to undertake something requiring an understanding of mathematics that many would think beyond the comprehension of secondary school students. I spontaneously replied that we were given puzzles at home on an almost weekly basis from an early age, unintentionally suggesting that this was the reason I didn't feel intimidated by mathematics. But, of course, this isn't entirely true: I'm intimidated by math the same way other people are, when they cannot make head or tail of what they're hearing or reading. It may be that this happens less often to me than to others as a result of thinking habits I acquired through puzzle-solving, but I have the same trouble as the other students whenever the math we're being taught is a little over our heads. The one thing I will say about Dad's giving of puzzles (which he continues to do to this day) is that by so doing neither he nor Mom prescribes boundaries on what they think we can or cannot do.

What's probably closer to explaining my supposed confidence is this: For years I had the advantage (though you might not call it that) of hearing mathematics spoken in a natural way between my father and a physicist friend of his in our kitchen. One of them would stand at the board writing math and explaining, while the other sat in a chair beside the table, listening and cross-examining him. Whenever I heard something like, "Surely, that can't be," I realized with some relief that the people I supposed could solve *every* math problem were, in fact, often struggling. I listened to their animated conversation as they discussed ideas, and if eventually they got the better of the problem, I enjoyed hearing them say things like, "I didn't know that was true." This was how I learned that you can talk about math without always getting to the answers, and that it wasn't all seriousness—that sometimes it even seemed to be fun. So when, through a combination of circumstances I'll explain in the next chapter, I became involved in cryptography, I never entertained the thought that someone my age could not possibly do this because it contains so much math. Instead, I embraced it with a sense of adventure, but safe in the knowledge that I could always call on Dad for help whenever I needed to—which I was sure to do.

3 Beginning My First Project

In October 1997, one of the science teachers at my school, Seán Foley, started to look for students who would be willing to do a project for the Young Scientist Exhibition. This annual competition is held in early January in Dublin. Almost everybody in the country knows about it, and each year there are hundreds of projects on display in the large hall at the Royal Dublin Society building. It began in 1964 and was sponsored by Aer Lingus, our national airline, until 1997, when Esat Telecom took it under its wing. The competition is now known as the Esat Young Scientist & Technology Exhibition. Its purpose is to encourage young people to become more involved in the sciences. It's a great idea, particularly if you are really interested in something, because you can investigate to your heart's content and do as much work as you want, knowing that you have the opportunity to present your ideas and the results of your labors in public. You may enter either with an individual project or, as part of a team of at most three, with a group project.

Mr. Foley, who had brought students to this competition for a number of years, told us that we'd find it very exciting and gain great experience by displaying and explaining our projects to judges and the general public. We would also have a tremendous time meeting and talking with the other participants. Mr. Foley said that there was great fun to be had at the social events organized for the evenings, and that we'd get to stay at Jury's Hotel for very little because the Parents Association would pick up the rest of the tab. It all sounded very interesting—the only trouble was I didn't have any idea for a suitable topic! I loved doing projects at school in subjects like geography and history, trying to come up with an unusual topic and include everything I could find. As one teacher commented, "You left no stone unturned and put a lot of effort into the writing." But those projects were really no more than gathering and marshalling information in an attractive way (in fact, organizing—which I love; I'm not quite as good at *keeping* organized, though . . .). I was conscious of this fact, so I was a little unsure about undertaking a science project with a practical component.

Still, I knew it would be worth doing something just for the *craic,* as we say, to be up in Dublin for five days. It was an ideal opportunity since I was in my transition year, with a lot of time that I could call my own.

In the Irish school system the transition year is an optional, nonexamination year between the end of the junior cycle (the first three years) and the start of the senior cycle (the last two years), designed to provide students with a year of wide-ranging study. Students who opt for the transition year cover the usual core subjects, but they also get to study noncore subjects like art, craft and design, desktop publishing, drama/public speaking, personal development, entrepreneurship and video production. On top of all this they get two weeks of work experience, which in my case became part of the reason I am now writing this book. But that's a story for later.

We had several collaborative projects during my transition year. We set up a company to make Christmas cards, Christmas logs and bookends, sold shares in it, marketed our product in advance and sold it at Christmastime for (I'm happy to say) a moderate profit—and then we learned how to wind down the company. We also put together a fashion show for which we went on a professional model training course. We organized the borrowing (and, later, the return) of very smart clothes from various shops in the city, then modeled the clothes ourselves at a show in front of all our teachers and the pupils from the other classes. It was a tremendous night. These were all great experiences, but I had the best fun of that year during a one-week visit to an outdoor education center, where we learned basic survival techniques, orienteering and rappelling, went canoeing and mountain climbing and, at night, had great sing-alongs.

Not all students opt to take transition year; some prefer to go straight on to the senior cycle so that they can leave school one year earlier. What's sad is that those who take the transition year risk losing the close contact they had built up with friends who take the shorter route, and vice versa. But for me this was the only unattractive feature of what was otherwise a most exciting and eventful year.

I decided to enter the Young Scientist competition, but I still had no clue of what to do for a project. One evening, as Dad and I were driving to the evening class I was attending and he was teaching (and of which I'll have *much* to say shortly), I asked him if he had any ideas. He knew about the competition because when he was at school many moons ago, a student one year ahead of him had won a prize and become a school hero overnight. He said he hadn't any ideas. He asked, "Won't you have to carry out experimental investigations, gather data and draw conclusions?" I explained some of the types of projects that had been submitted in more recent contests, which Mr. Foley had described to us when he was trying to sell us on entering the competition.

After many "I don't knows" and as many hems and haws, Dad said, "Maybe you could do something on cryptography—perhaps discuss some of the ideas, and then for the 'bells and whistles' end, do some programming. You could show messages being scrambled* to disguise them, and then unscrambled† by the receiver to recover the original message." He thought it might be a good idea to have all this happening between two PCs.

I had never done any programming, but I knew what it involved because I had heard Dad rave about a program called *Mathematica*. We talked a little about the fact that most people don't know anything about public key cryptography, which dates from 1976, and we thought they might find the subject very in-

*Enciphered or encrypted.

† Deciphered or decrypted, but you probably figured that out already.

teresting, particularly if I could show that these revolutionary ideas are of practical importance. Hmm, maybe. By this time Dad's evening class had covered a good deal of number theory* and we were discussing cryptography, so if I did decide to take the plunge I'd have an extra reason to pay attention. Besides, it would be an opportunity to learn what the "wonderful piece of software" *Mathematica* could do, and I had an inkling that I would really love programming.

On the downside, I knew there had been some mathematical projects entered in previous competitions, but I believed they'd been few and far between. I had some notion that the format for competition entry was not ideally suited to abstract projects, and was more suited to ones involving experimental work or field-work, which seemed quite natural. Even so, there is a category called "Physics, Chemistry and Mathematics," and, after all, I'd be presenting the *programming* of the mathematics as the practical element. Why not give it a shot?

That's how I got started on what I now call my first project, which I simply entitled "Cryptography—The Science of Secrecy." My plan was to present the terminology and notions by using as examples some of the simpler classical cryptosystems,† beginning with the Caesar cipher. This system is very easy to describe and understand, and ideal for introducing all the key terms. I would then present the fascinating ideas of public key cryptography, culminating in a description of the celebrated RSA system. All this would be accompanied by programs showing cryptosystems in action. Later, I made a wheel to illustrate the Caesar cipher. Just making it was fun. The wheel looked great all prettied up, and it proved to be a big hit because it illustrates certain ideas which everyone can understand.

I knew I would have to do a lot of work, but as usual I thought I could leave everything to the last minute, maybe a

*Just what it sounds like—the study of the properties of whole numbers.

† A cryptosystem is a way to mask messages so only intended parties will know what's behind the mask.

couple of weeks or so before the competition, because I like working under pressure. But Dad encouraged me to get cracking as soon as possible: "It'll take time to assimilate the concepts, to let them mature and become clear in your mind. Anyway, you are almost sure to run into all sorts of difficulties at the programming end when you try to implement the mathematics. Projects don't have to be hectic affairs; better if you give yourself plenty of time and that you take your time." I knew he was right.

He was dead right about the programming. Near the end, when it came to writing up my report, I realized just how much work is involved in trying to present ideas clearly and simply without being overly technical. It is difficult trying to be crystal clear but well worth the effort if you succeed. I was to learn this lesson most emphatically when writing programs: you must know exactly what you want to achieve and write the code accordingly, otherwise it won't do what you hope it will and you end up tearing the hair out of your head. When you have to explain things clearly you really have to think from the bottom up. Wondering, "How am I going to explain this or that to a layperson?" and asking, "Well, what is really involved?" are important. Dad has some quote (he always has!) from a French mathematician that goes: "A mathematical theory is not to be considered complete until you have made it so clear that you can explain it to the first man whom you meet on the street."* This is a pretty tall order and it may not always be achievable. More comforting is something I heard attributed to Albert Einstein, which I paraphrase as, "You should strive to make things as simple as possible but no simpler." Now that's a little more realistic, as well as a witty warning against oversimplification.

If you really want to learn something, learn it with a view to explaining it to somebody else. I found Mom to be the best person

*Please don't ask me *which* French mathematician. The German mathematician David Hilbert said in 1900 that it was "an old French mathematician," and that's good enough for me!

on whom to test this theory. She thinks slowly and carefully and will let you get away with absolutely nothing. At times you might think her dense, until you realize that you're the dense one for expecting her to grasp some idea you've presented to her without having supplied her with vital facts. So I learned to order my thoughts. She asked tough questions which always sounded simple but were right to the point.

On a car trip to Dublin, I was trying, by way of practice, to explain some cryptographic ideas to her. I'd talk for a while and then she'd ask a question. Dad, who was snoozing in the front passenger seat, would pipe up every now and then with "That's a great question." I felt a bit small on hearing this because perhaps I hadn't asked myself that question. But Dad didn't mean to make me feel this way, though when it comes to our explaining things he doesn't spare our feelings if he thinks we're waffling. He often advises, "Speak clearly and simply. Tell a person what you know and come clean when you don't know something. Don't bluff." I know his general attitude is that you should ask simple natural questions of yourself—put yourself in the other fellow's shoes and try to see his difficulties. At one point while I was explaining to Mom something called the RSA signature scheme (of which I'll talk later), she went, "Let me get this clear. You are saying that you actually decrypt first and then encrypt." I'll always remember that remark. It was right on the nail, and I could see her difficulty—how could you be decrypting ordinary text (plaintext), and why would you be doing it? I had to think for a while before I could explain. This explaining business really kept me on my toes. Dad loved it all, and again said, "They're great questions, Mom." At this Michael, who by this time was completely frustrated at being cooped up in a car listening to all this crazy talk, saw his opportunity and struck: "Dad thinks every question is a great question! 'What time is it?' is a great question in Dad's mind." There was general laughter and no more cryptographic talk. We took a break and enjoyed the scenery, even though it was probably raining.

Learning the Relevant Mathematics

The project required that I learn number theory (really learn it), the necessary cryptographic ideas *and* how to program in *Mathematica* so that I could implement and illustrate the cryptographic schemes. The first thing I realized about learning mathematics was that there is a hell of a difference between, on the one hand, listening to math being talked about by somebody else and thinking that you are understanding, and, on the other, thinking about math and understanding it yourself and talking about it to somebody else.

In the beginning I was inclined to accept mathematical facts in the same way that I accept geographical ones. I felt that I grasped concepts fairly quickly, but I now know that I didn't ask *why* things are the way they are often enough, as a true scientist should. As a result, I found that sometimes I didn't appreciate how important a certain detail might be. Knowing that Paris is the capital of France is something you don't think much more about, but to know what Fermat's Little Theorem (which I'll tell you about later) is saying requires thought. You really have to tease out under what conditions the result is true, and explore it to get to know what exactly is being asserted. This was all a little intimidating at first, but I always knew I could call on Dad if it got too much for me. He's only too willing to talk math all day long.

I was very lucky in that I quickly found programming to be a great help in learning some of the mathematics. Many results in number theory are ideally suited to being programmed, and the mere act of trying to write a few instructions in sequence to display some examples of a general result helped me learn this type of mathematics in an easy and natural way. I even got to appreciate some of those details that many often find fussy or annoying in mathematical results. From my perspective, overlooking some significant little detail, while frustrating, was nothing to get too het up about. It was like someone wondering why the car won't start only to find that the reason is as simple as having forgotten to put petrol in the tank. Because of this I became less inclined to think,

Why does this result have to be so restrictive in what it says? I could see the sense of the details, so I learned to be tolerant.

As planned, I used *Mathematica* to implement the cryptographic schemes. This program allows you to do the same sorts of computations you might do with a calculator, but with much larger numbers. More importantly, you can also use it to write programs. That was something I was going to have to learn. For me the initial thrill was simply seeing how this package deals with huge numbers with such speed. I just love to calculate—even the simplest of calculations (getting it to print the first 100 digits of π or expand 3^{76} so that I could see the numbers roll across the screen). It is spectacular to think how long it would take to do these by hand. Contrast the time people spent working out successive digits of π (and making up rhymes so that they could remember them!) with the microseconds *Mathematica* takes to reel off the first thousand digits of this number. Learning simple commands such as **PrimeQ[ ]** or **FactorInteger[ ]**, which aren't on a calculator, made me feel powerful, though in reality I wasn't doing anything. Typing **PrimeQ**[2^{32} + 1] and getting the output **False** as soon as the command was entered was wonderful, especially when I remembered that the great Fermat believed this number to be prime. Typing **FactorInteger**[2^{32} + 1] to get in the blink of an eye the factors 641 and 6,700,417 was even more exciting.

After I spent much time playing with the many different commands (some of which I knew I would never need), I decided it was time to take my first baby steps in programming. One of the first tasks I set myself was to go back through my number theory notes and search out simple numerical statements which would involve huge amounts of labor for a human to illustrate by examples, but which would be no bother for a computer. So I began my programming by writing short sequences of instructions to carry out the necessary calculations. For example, I wrote code to make out lists of the Fermat, Euler and Mersenne* numbers along with their factorizations. This was very straightforward, and the results were printed in

*We'll hear more about these gentlemen and their numbers very soon.

a matter of seconds. Then I tried to fine-tune the code a little to get more information, such as which ones are prime and which are not. This was a little more challenging. I found that as I worked away, improving here and there, other little refinements would suggest themselves. It wasn't always easy to tell the computer how to carry them out. I acquired a great respect for how versatile the human brain is.

When I grew in confidence I started to learn how to turn ordinary text (plaintext) into scrambled text (ciphertext), a process known as *enciphering*. This wasn't too hard, but I did spend a lot of time reading the relevant sections of the 1500-page manual, all the while fighting the temptation to explore some of the wonderful new commands I came across as I leafed through this massive tome. Eventually I got to the stage where I could turn chunks of plaintext into huge numbers, which I then turned into different huge numbers, which then got converted into blocks of ciphertext. The next stage was to reverse the process, which is called *deciphering*. Although the two processes are the same mathematically, it was for me much more of a test because I may not have been doing the enciphering properly. So I had my fingers crossed that all would work with the reversing process, nervously wondering, will the original plaintext reappear? What joy when it did! I knew it should, but it was still such a relief to see the correct plaintext pop up on the screen, verifying that the code I'd written, however primitive it might be, was doing its job.

Often one of those annoying little bugs would have me tracking it for hours, forcing me to break up entire sections of code, going down through it step-by-step to pinpoint exactly where the little pest was hiding and how it was managing to mess up what otherwise seemed a perfectly logical program. Line by line I'd dissect and examine code. Yes, this is working, so is that. Now what is happening here? Try this—no good. Whoops—I've made it worse. "Erase and rewind 'cause I'm a-changin' my mind," as the song says. Maybe it is this—Noooo! (followed by foul language). OK, relax! Don't be in such a rush. Think. Think. What am I asking the program to do? Yes. Is that what I meant it to do? Oops, not quite! Ah yes, it is doing that, but I wanted it to do this. Gotcha!

Part II

Mathematical Excursions

4 Dad's Evening Class

God made the natural numbers, all the rest is the work of Man.

—LEOPOLD KRONECKER

These were the first words that my father wrote on the board—a bit alarmingly, in German—on the evening of my first class at the Cork Institute of Technology, where he lectures in mathematics. He explained that Kronecker was a German mathematician who lived in the nineteenth century. Little did I know the role that the natural numbers (1, 2, 3, and so on . . .) would play in my life over the next two years.

Although, as I have already said, my father had often taught or helped with my school mathematics at home, this was the first time I had ever attended a formal lecture of his and seen him present work in an organized way to a group of people. This unusual evening class had come about as a result of a conversation he and I had the previous summer, which started something like, "Now that you have decided to do transition year, I must do some math with you." He continued, "I'd like to show you how some beautiful but reasonably elementary mathematics is applied, stuff that you wouldn't ordinarily come across in school."

Not knowing exactly what he had in mind, but not wishing to be collared and dragged to the kitchen blackboard at random times, I replied, "Dad, whatever you do, do something structured!" He promised to think carefully about the most effective but gentle way to give me a realistic glimpse into the mathematical world. Conscious of how much *he* would have appreciated it had someone done this for him thirty years ago, he felt obliged now to offer me the benefit of his mathematical knowledge. "Of course, only if you are genuinely interested—I wouldn't force it on you."

Of course he wouldn't! But I *was* interested. So this is how the evening class was conceived, and this is how I was to be introduced to many wonderful mathematical ideas—accessible ideas I might never otherwise have encountered unless I chose to pursue mathematics as a career.

The title for the intended series of lectures, "Mathematical Excursions," was suggested by one of Dad's favorite books, *Excursions in Calculus* by Robert M. Young. Dad wanted not to present mathematics in a very rigorous fashion, but rather to explore "elementary" aspects of "higher" mathematics, and learn how these ideas find applications in different fields of study outside mathematics. The only prerequisite for these excursions—besides a thorough grounding in ordinary arithmetic and basic algebra—was interest. The primary aim of the course was that everybody take pleasure in sharing in the spirit of discovery.

The class, which met Tuesday evenings from seven to ten, consisted of eight people and ran for twenty-five nights. At fifteen and a half I was the youngest, though there were a couple others nearly as young. The rest were adults of various backgrounds: one was a secondary school mathematics teacher, one a chemistry graduate working in a medical laboratory, and others were computer scientists. All were there either because they had loved math at school but hadn't continued with it, or because they felt they hadn't properly appreciated the subject the first time around. This noncredit class, which was not part of any major, led to no award or certification at its completion and demanded no homework or study, may have seemed an ideal way of "getting back into math" in an easygoing, informal manner. Dad told us at the outset that there would be no pressure to perform, and that we need not fear being asked questions but we could ask them at any time. We were to feel free to say "any crazy thing" we liked, and not to be in the least afraid of appearing to make fools of ourselves. In fact, he told us, the more you are prepared to make a fool of yourself the more you'll learn. These reassurances convinced me that I was going to enjoy the next few Tuesday nights, even if I was not completely comfortable with the fact that the lecturer was my father. (Students always pity those who have to endure classes given by their parents, though I'm sure the situation is worse for the parent who is the teacher.)

The first excursions were to be in elementary number theory, with some aspects of cryptography as their final destination. Along the way we would see interesting sights, rich in their own right regardless of whether or not they had immediate application. Dad told us that when he was a student, number theory was regarded as the sole province of mathematicians, and those who devoted their lives to its study were considered the purest of the pure, so few were its perceived applications. It wasn't until the 1960s that engineers and scientists began to find applications of number theory, and by the end of the 1970s the discipline came of age, in the applied sense, with the advent of public key cryptography (very briefly, a modern form of cryptography which makes it possible to make public the method by which messages are enciphered, but—amazingly—*without* revealing how these enciphered messages are *de*ciphered). Now the U.S. government was putting millions of dollars into the study of cryptography, a subject full of wonderful ideas about which, Dad said, he could hardly wait to tell us. Whereas many great number theorists had barely made a living in the past, their modern counterparts were now being actively sought out, so great was the demand for expertise in this emerging science of secrecy which relied so fundamentally on the branch of mathematics that the great mathematician Karl Friedrich Gauss (of whom more later) revered above all else. It was he who said:

> Mathematics is the queen of the sciences and number theory is the queen of mathematics.

As you might imagine, hearing all this filled me with curiosity and interest. Mom said it was like the Sleeping Beauty fairy tale, and she was prompted to write:

> A princess was the Theory of Number,
> whom no practical use did encumber,
> 'til Cryptography the prince
> ('tis all secrecy since),
> did kiss her awake from her slumber.

I thought about those who had toiled away through the centuries at unraveling the mysteries of this subject, motivated by nothing more than a passionate desire to know. They could have hardly dreamt of the applications that some of their results would one day find. I wondered what it was they had discovered, and what they would think if they could see how some of these discoveries are now being used. I was eager to learn the subject and surmise for myself whether they would be surprised or not.

We were at once tantalized and put on our guard by the warning that number theory is notoriously deceptive. We were told that very soon we'd find ourselves asking *simple* questions, many of which others have asked before us through the ages, and some of which have still not been answered to this day.

5 Of Prime Importance

Let's begin with the natural numbers, those seemingly simple things which Dad and Herr Kronecker found so important.

A natural number is divisible by another natural number if and only if the first number is a "natural" multiple of the second one. Thus 98 is divisible by 7 because when 7 is multiplied by 14 the result is 98. In this case the number 7 is said to be a factor, or a divisor, of 98. By the same token, 14 is also a factor, or divisor, of 98. Because $98 = 98 \times 1$, the numbers 1 and 98 are also factors of 98, but they are termed the trivial, or improper, factors of 98 because they're obvious and no labor is required to find them. On the other hand, the factors 7 and 14 are termed nontrivial, or proper, factors of 98.

We often say that 98 is the product of its factors 7 and 14, while we refer to 7×14 as a proper factorization of 98. An improper, or trivial, factorization is 1×98 (which is the same as 98×1). Two other proper factors of 98 are 2 and 49, with 2 being the smallest nontrivial factor of 98, and 49 the largest. Thus 2×49 is another proper factorization of 98.

The natural number 99 is not divisible by 7 because there is no natural number which when multiplied by 7 gives 99. Thus 7 is not a factor, or divisor, of 99. However, 3 is a factor of 99, as is 11, but whereas 11 is termed a single factor of 99, the factor 3 is a repeated factor because $3 \times 3 = 9$ is also a factor of 99.

Every natural number greater than 1 is always divisible by itself and 1, and so always has at least two factors and a trivial factorization. However, among the natural numbers there are many which have more than just the two trivial factors. As we have just seen, 98 is one such number. Another example is the number 12, which has exactly six factors. They are 1, 2, 3, 4, 6 and 12; the four factors 2, 3, 4 and 6 are the proper factors of 12, while 1 and 12 are its improper, or trivial, factors. The fact that so many lesser numbers divide "evenly" into 12 is one reason the old English shilling was equal to 12 pennies. Here is a little puzzle that might get you thinking about factors and factorization.

Can you find the number less than 1000 which has the most factors? I won't tell you the answer, but I will tell you that there is only one and that it has exactly 32 factors. If this strikes you as a little too much like work, or not particularly interesting, don't worry—just read on. (But if you'd like to know more right now, see Appendix B, page 297.)

There are also natural numbers that have no more than the two trivial factors. Twelve of the first thirty-eight natural numbers, namely

2, 3, 5, 7, 11, 13, 17, 19, 23, 29, 31, 37

have exactly two factors. These numbers are the first dozen of what are known as the prime numbers. A prime number, or prime for short, is a natural number which has only two factors, itself and 1. Thus a prime cannot be written as the product of numbers smaller than itself. As we shall see, the primes are the elite of the natural numbers.

Those numbers that have more than two factors are known as composite numbers. Thus 12 and 98 are composite numbers. We can make a geometric distinction between prime numbers and composite numbers. A row of 12 coins

● ● ● ● ● ● ● ● ● ● ● ●

can be arranged into a rectangular array consisting of three rows each with exactly four coins:

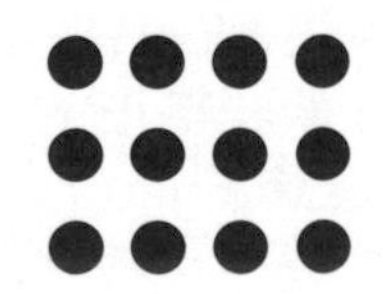

This is possible because 12 is composite (short for composite number) and so has a nontrivial factorization given by $12 = 4 \times 3$.

This row of 12 coins can also be arranged as

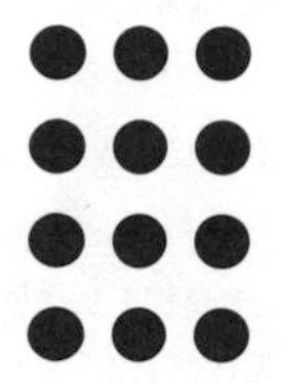

which consists of three rows of four, and reflects the fact that $12 = 3 \times 4$. Note that this arrangement is the previous one rotated through 90° so that the rows become columns, and vice versa.

Another arrangement is

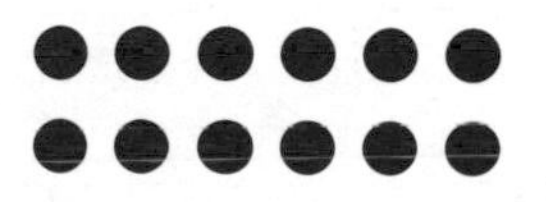

which may be thought of as two rows with 6 coins each or six columns with 2 coins each. The rectangle is a geometric illustration of the fact that $12 = 2 \times 6 = 6 \times 2$.

On the other hand, a row consisting of a prime number of coins can never be rearranged into a rectangular array with more than one row where each row has the same number of coins. Add just one more coin to the row of 12 coins above to get a row of 13 coins

and, try as you will, you will never succeed in rearranging this single row into a rectangular array of more than one row with each row having the same number of coins.

You might think that because a number is "big" it is bound to have lots of factors, but surprisingly this is not always true. The number 889,363, which is not far from a million, has only its two trivial factors. It has no proper factors. A single row of 889,363 soldiers could never be rearranged into ranks and files so that each file contains the same number of soldiers. Thus 889,363 is another example of a prime. If you just add one more soldier, you *can* form a rectangle consisting of 644 rows each with 1381 soldiers, or a rectangle with 161 rows each with 5524 soldiers. Are there other rectangular arrangements? (To verify your answer, see Appendix B, page 298.)

The natural number 2 is the only even prime number and so is considered the *oddest* prime of all. Pairs, such as {3,5}, {5,7} and {11,13}, in which the larger prime of the pair is greater than the smaller one by exactly 2 are termed *twin primes*.

Now that we know the difference between a prime number and a composite number we can ask all sorts of questions, two of which might be:

- How do you factor a number?
- How many prime numbers are there?

You might think it's easy to answer the first question, but it is, in fact, one of the most difficult and important problems in number theory. This problem is absolutely central to the main body of modern cryptography, which would crumble if someone came up with an efficient way of factoring an arbitrary "large" number. The second question, which may strike you as harder, was answered over two thousand years ago.

So you see that although we have barely begun to talk about numbers, we are already posing questions which appear simple but whose answers are far from immediate, and may even elude us. As Paul Erdös, the twentieth-century Hungarian mathematician known as the Magician from Budapest, remarked: "Babies can ask questions about primes which grown men cannot answer."

Some other "babylike" questions are:

- How do you decide whether or not a number is prime?
- How many prime numbers are there less than a given number?

- Is there a formula for generating prime numbers?
- Is there an infinite number of twin primes?

The assertion that the answer to the last question is yes is known as the twin prime conjecture because nobody has been able to pronounce on it yet one way or the other. As the American mathematician (and lawyer!) Peter Rosenthal wrote,

> The infinitude of the primes
> Is the subject of plenty of rhymes,
> But we can't begin
> To prove there's a twin
> An infinite number of times.

The number 420 is composite because it can be written as 10×42. Another nontrivial factorization which also reveals 420 to be composite is 14×30. This factorization may *look* as if it has nothing in common with the previous one and so might lead one to believe that 420 can be constructed in different ways. However, neither of these apparently different factorizations is *complete* because, in each case, the factors in the products can themselves be "broken down," or decomposed, by further factoring. A factorization is complete only when all the factors appearing in the final product are primes. This is because primes cannot be factored nontrivially. Primes are to atoms what composites are to molecules. Doing this for the number 420, starting with each of the incomplete factorizations given above, we get

$$\begin{aligned} 420 &= 10 \times 42 \\ &= (2 \times 5) \times (3 \times 14) \\ &= 2 \times 5 \times 3 \times (2 \times 7) \end{aligned}$$

and

$$\begin{aligned} 420 &= 14 \times 30 \\ &= (2 \times 7) \times (5 \times 6) \\ &= 2 \times 7 \times 5 \times (2 \times 3) \end{aligned}$$

Both these combinations of primes, when arranged in order of increasing magnitude, give

$$420 = 2 \times 2 \times 3 \times 5 \times 7$$

Here 420 is represented in a unique way as a product of primes. Had we begun with $420 = 15 \times 28$, we would still have arrived at this same *prime decomposition* of 420.

Every natural number has a unique prime decomposition, which is something that may seem self-evident but is, in fact, far from obvious. The uniqueness of the complete factorization of an integer is something very fundamental which had to be proved. It is known as the *fundamental theorem of arithmetic,* or the *unique factorization theorem;* simply put, any positive integer can be represented in exactly one way as a product of primes.

Since every natural number is made from a unique combination of prime numbers, we may think of the primes as the building blocks for all the numbers. I must admit that I wondered how things could be otherwise. But who can say without proof that the number

$$654{,}298{,}765{,}432{,}987{,}654{,}321{,}111{,}398{,}503{,}400{,}542{,}111$$

can be decomposed in one way only? Isn't it conceivable that two or more different combinations of primes might, when multiplied out, give this answer? The fact that this cannot happen is very important when it comes to cracking cryptographic codes.

To help you appreciate fully this unique factorization property of the natural numbers, let me tell you about the *Hilbert numbers,* named for David Hilbert. The first few Hilbert numbers are

$$1, 5, 9, 13, 17, 21, 25, 29, 33, 37, 41, 45, 49, 53, 57, \ldots$$

Apart from 1, each of these numbers is 4 times a natural number plus 1 (e.g., $45 = 4 \times 11 + 1$). If you multiply any two of these numbers you always get another one of the same type. (This is not hard to show using a little basic algebra.) Focusing solely on these numbers—and imagining that no other numbers such as 3 and 7 exist—we can, as with the natural numbers, talk about Hilbert primes such as 5, 9, 13 and 17 (no one of these is divisible by a smaller Hilbert number other than 1) and Hilbert composites such as 45 (which is the product of the two Hilbert primes 5 and 9). Now comes a surprise. The calculation

$$1617 = 21 \times 77 = 33 \times 49$$

shows that the Hilbert number 1617 can be completely factored as a product of Hilbert primes in two different ways. The numbers 21, 33, 49 and 77 are all Hilbert primes (though none of them is a natural prime).

Because of this lack of unique factorization, arithmetic with Hilbert numbers might prove a tad more complicated, if not more interesting. Maybe we should use this arithmetic to confuse our enemies. "Which two Hilbert primes did I multiply to get 1617?" Whichever way they answer we can say, "Sorry, wrong," and give the other combination.

How Do You Decide Whether or Not a Number Is Prime?

Let's imagine a conversation between us. Note what clever questions you ask!

YOU: How do you decide whether or not a natural number is prime?

ME: Simple. Step I: check whether it is divisible by 2. If it is, it is composite (unless, of course, it's 2, but come on, let's be serious). If it is not, proceed to Step II: check whether it is divisible by 3. If it is, it is composite. If not, proceed to Step III: check whether it is divisible by 5.

YOU: Hold on a second! You've just skipped 4. What about checking whether it is divisible by 4?

ME: There's no need. If it's not divisible by 2, then it certainly is not divisible by 4. If it's divisible by 5, then it's composite; otherwise proceed . . .

YOU: Don't tell me. Step IV: if it's divisible by 3 it's divisible by 6, so check whether it's divisible by 7, the next prime after 5.

ME: That's right.

YOU: But this checking to see whether the number is divisible by primes could go on for a long time if the number is large. And wouldn't I need to know a lot of primes?

ME: You do need to have a store of primes for the purposes of this trial division procedure. But leaving that problem aside for the moment, how many tests would you have to carry out to determine whether or not 149 is prime?

YOU: You might have to divide 149 by all the primes up to 149.

ME: That would certainly tell you whether 149 is prime, but the good news is that there is no need to do this much work. Ask yourself, Would 23 be the first prime factor of 149?

YOU: Why not?

ME: Well, if 23 were the first factor then the other factor would have to be at least 23. But $23 \times 23 = 529$, which is much bigger than 149.

YOU: Oh! I see. So how far up do you have to test?

ME: If 149 has factors, they cannot all be greater than its square root.

YOU: So you have to go up as far as the square root of the number.

ME: Yes. Now, $\sqrt{149} = 12.2\ldots$, so if 149 is composite it must have a prime factor less than 13. So I need only test to see whether it's divisible by the primes 2, 3, 5, 7 and 11. At most I need to do five trial divisions to decide whether or not 149 is prime.

And since it's not divisible by any of these five primes,

the number 149 must also be a prime. You can add this to your collection of prime numbers and use it as a trial divisor in other calculations testing numbers that are larger than $149^2 = 22{,}201$.

YOU: So I can use this procedure either to find a prime factor of the number I'm testing or establish that the number itself is a prime.

ME: Yes. This procedure is called the trial division algorithm. "Algorithm" is just a fancy word for a step-by-step procedure.

YOU: That's great. With superfast computers you should be able to test any number fairly quickly.

ME: Unfortunately, it's practical to use this algorithm only on reasonably small numbers. That's the bad news about this wonderfully simple procedure. Even if you avoid storing huge amounts of primes in the computer's memory by getting the computer to divide the number being tested by every number up to its square root, it can take a very long time for numbers with as few as forty digits. Let me quote the number theorist Carl Pomerance:

> The main drawback of the trial division algorithm is that it takes too long if a number has no small prime factors. For example, say we use the trial division algorithm on a computer that can do one million trial divisions per second to determine if a given number is prime or composite. If the number is a prime near 10^{40} (a number with approximately 40 digits), the running time will be about one million years while if it's a prime near 10^{50} (a number with approximately 50 digits), the age of the universe would not suffice.

That's pretty bleak. Although in theory it is simple to decide whether or not a natural number is prime, this method is not practical. Luckily, as I shall mention much later, there's another method that is.

The Sieve of Eratosthenes

Around 250 BC the Greek mathematician Eratosthenes used a spectacularly simple method to find primes which has become known as the sieve of Eratosthenes. Here is how it finds all the prime numbers among the first 120 numbers, which are listed, with the exception of the number 1, in the following array:

–	**2**	3	4	5	6	7	8	9	10
11	12	13	14	15	16	17	18	19	20
21	22	23	24	25	26	27	28	29	30
31	32	33	34	35	36	37	38	39	40
41	42	43	44	45	46	47	48	49	50
51	52	53	54	55	56	57	58	59	60
61	62	63	64	65	66	67	68	69	70
71	72	73	74	75	76	77	78	79	80
81	82	83	84	85	86	87	88	89	90
91	92	93	94	95	96	97	98	99	100
101	102	103	104	105	106	107	108	109	110
111	112	113	114	115	116	117	118	119	120

The even numbers 4, 6, 8, 10, . . . are all composite since their respective factorizations are 2×2, 2×3, 2×4, 2×5, and so on. The first sieving, "based on the prime" 2, proceeds as follows: beginning with the number 4, "sieve," or delete, every second number by leaving the corresponding cell blank. This operation of skipping over numbers, or cells, to reach other ones involves nothing more than counting. When this is done the original array becomes

–	**2**	3		5		7		9	
11		13		15		17		19	
21		23		25		27		29	
31		33		35		37		39	
41		43		45		47		49	
51		53		55		57		59	
61		63		65		67		69	
71		73		75		77		79	
81		83		85		87		89	
91		93		95		97		99	
101		103		105		107		109	
111		113		115		117		119	

Since the number 3 has not been sieved, it is the next prime and is used for the second sieving. Now empty every third cell after 3: one, two, delete, one, two, delete, and so on. It doesn't matter if a cell, such as that corresponding to 6, is already blank. The reason blank cells (corresponding to numbers that have already been sieved out) are never removed is to ensure that the skipping proceeds correctly at all stages. The blank cells are a vital record of where numbers were. After this second sieving, the array becomes

–	**2**	**3**		5		7			
11		13				17		19	
		23		25				29	
31				35		37			
41		43				47		49	
		53		55				59	
61				65		67			
71		73				77		79	
		83		85				89	
91				95		97			
101		103				107		109	
		113		115				119	

It is worth noting that the first number to be sieved by this second sieving based on 3 is its square, $9 = 3^2 = 3 \times 3$.

Since the number 5 has not been sieved by either of the two previous sievings, it must be the third prime and the basis for the next sieving. Leaving every fifth cell after 5 blank, the following array emerges:

–	**2**	**3**		**5**		7			
11		13				17		19	
		23						29	
31						37			
41		43				47		49	
		53						59	
61						67			
71		73				77		79	
		83						89	
91						97			
101		103				107		109	
		113						119	

This time the first number to bite the dust, having survived the previous two sievings, is $25 = 5^2 = 5 \times 5$. Now 7 is the first number appearing after 2, 3 and 5 in this latest array. Since it has survived all three previous sievings, it must be the next prime. Leaving every seventh cell after 7 blank gives the following array:

–	**2**	**3**		**5**		**7**			
11		13				17		19	
		23						29	
31						37			
41		43				47			
		53						59	
61						67			
71		73						79	
		83						89	
						97			
101		103				107		109	
		113							

Again note that the first new number deleted when skipping in sevens is 7^2, or 49. Now stop. Why? Because the next sieving, based on the prime 11, crosses out every eleventh number after 11. The first multiple of 11 which isn't already sieved is the eleventh; the lower multiples of 11 have already been deleted by those sievings based on one or other of the primes 2, 3, 5 and 7. Hence 121 (11^2) is the next natural number for the chop, so those thirty numbers less than 121 clearly visible in the last array are in no further danger. These survivors are the thirty primes less than 120. It took just four sievings to reveal them. (Do you now see why the first 120 numbers were chosen to demonstrate this remarkably simple algorithm? If not, check Appendix B, page 298.)

Although the sieve of Eratosthenes may look like a very laborious method, it is, in fact, a very fast procedure since, by contrast with the method of trial division, it does not make use of time-consuming divisions. It is a very nice programming exercise to get this working and fine-tuned. When you do get it going, it works like a charm and lists all the primes up to a given number in no time. It will fill the screen with the 1229 primes less than 10,000 before you have time to catch your breath. The fact that its output can be huge is something of a disadvantage, but I'm told that this simple principle is the basis of a very powerful modern method used in the constant quest for larger and larger primes.

I was to learn a lot about the difference between theory and practice during my first project in cryptography. I would come to appreciate that it wasn't good enough to know any old way of doing something—you have to find methods or techniques that implement the theory in highly efficient ways. This "real-life" aspect caught my interest. Great theory, but is it implementable? It's no use receiving a secret message scrambled in a highly secure way which says, "Evacuate immediately!" if it is going to take a day to work out what it is saying.

But there is some good news. The amazing thing is that there *are* other ways of figuring out if a number is prime, which work quite speedily on large numbers. But this is a story for later.

Is There an Infinite Number of Prime Numbers?

Although the ancients were able to find many primes with the sieve method, and today we actually know millions upon millions of primes, could they run out? This would mean that there is a largest prime number, after which all the remaining numbers stretching to infinity are composite. So is there only a finite number of primes, or is there an infinite number of them? I remember thinking how you would go about answering questions such as these. Do they necessarily have answers? I know that logically you can say that there is either an infinite number of primes or there isn't; it must be one or the other, but maybe it is not within our power to say which is the case. Subconsciously I wondered how people long since dead could have found answers to such questions.

Over two thousand years ago the Greek mathematician Euclid, who lived in Alexandria, Egypt, published in the sixth volume of his famous book on geometry a wonderfully simple proof that there is an infinite number of primes. You can find this proof in textbooks on number theory, but you can also find an excellent explanation of this beautiful result in Simon Singh's book *Fermat's Last Theorem.* (If you don't have a copy, you should get one because it's a great book.) Euclid's proof is an example of a "proof by contradiction." He begins by assuming that there is only a finite number of primes and then, by a simple but clever argument, shows that this assumption leads to a contradiction. Consequently there must be an infinite number of primes. It is a wonderful nonconstructive proof: it does not construct or exhibit an infinite set of primes explicitly, it just convinces you that there has to be an infinite number of them. It is an example of an "existence proof." It is similar to proving that there are at least two people in Dublin with the same number of hairs on their head, without actually producing two such people. You may know with certainty that they are out there somewhere, but please don't ask me to find them! (See Appendix B, page 298.)

Is There a Formula for Generating Prime Numbers?

Knowing that there is an infinite number of primes is different from knowing how to find any of them. Some gallant attempts have been made to find formulae that generate prime numbers only. Pierre Fermat (1601–1665), the great French mathematician renowned for his Last Theorem, mistakenly believed that each time a natural number n is substituted in the formula

$$F(n) = 2^{2^n} + 1$$

a prime is generated. Here $F(n)$ does not mean F multiplied by n, which would be written $F \times n$ or Fn. Read $F(n)$ as "the nth Fermat number." Then $F(1)$ is the first Fermat number, $F(2)$ is the second Fermat number, and so on. Fermat wasn't asserting that this formula would supply all the prime numbers, only that it would always produce a prime number each time a different natural number n is substituted into the formula. Of course, if he could have proved his assertion he would have had a proof different from Euclid's that there is an infinite number of primes.

Substituting 1 for n in the above formula gives

$$F(1) = 2^{2^1} + 1 = 2^2 + 1 = 4 + 1 = 5$$

So the first Fermat number, $F(1)$, is 5, which is a prime. The Fermat numbers obtained when 1, 2, 3 and 4 are substituted for n in turn in the above formula (note that 2^{2^3} means $2^{(2^3)}$ and not $(2^2)^3$) are

$$F(1) = 2^{2^1} + 1 = 2^2 + 1 = 4 + 1 = 5$$
$$F(2) = 2^{2^2} + 1 = 2^4 + 1 = 16 + 1 = 17$$
$$F(3) = 2^{2^3} + 1 = 2^8 + 1 = 256 + 1 = 257$$
$$F(4) = 2^{2^4} + 1 = 2^{16} + 1 = 65{,}536 + 1 = 65{,}537$$

These Fermat numbers are all prime. Now, $n = 5$ and $n = 6$ generate

$$F(5) = 2^{2^5} + 1 = 2^{32} + 1 = 4{,}294{,}967{,}297$$
$$F(6) = 2^{2^6} + 1 = 2^{64} + 1 = 18{,}446{,}744{,}073{,}709{,}551{,}617$$

The first of these is a ten-digit number which Fermat believed to be prime, and the second is a twenty-digit number which was well beyond the ability of anybody to factor in the 1640s, or, in fact, at any time before the advent of computers in the twentieth century. However, it is surprising that he did not discover that 641 is a factor of $F(5) = 4{,}294{,}967{,}297$ because when challenged by his compatriot Father Mersenne (1588–1648) to factor the larger twelve-digit number 100,895,598,169, he gave the factorization $112{,}303 \times 898{,}423$. The smaller factor, 112,303, of the twelve-digit number is much larger than 641. (The number 112,303 is the 10,650th prime, while 641 is only the 116th prime.) Perhaps Fermat didn't try very hard. If he mistakenly bypassed 641 when testing $F(5) = 4{,}294{,}967{,}297$ to see whether it is a prime, then it is understandable why he would have pronounced $F(5)$ to be prime. Its one remaining factor is the seven-digit prime 6,700,417 (the 457,523rd prime).

It was the Swiss mathematician Leonhard Euler (1707–1783) who, almost a century later in 1732, found the prime factors 641 and 6,700,417 of $F(5)$. The fact that it was nearly a hundred years before someone found how to factor $F(5)$ is an indication that factoring may not, in general, be an easy task. (Of course, you might say that nobody could be bothered to try.)

What about the number $F(6) = 18{,}446{,}744{,}073{,}709{,}551{,}617$, or Fermat-six, as it is now known? It is not a prime. Eighty-two-year-old F. Landry found its factors by trial division in 1867. So that you can appreciate what an impressive feat this was, imagine that you are thrown into prison and are told that you'll get out when you find the factorization. How long a sentence do you think you'd serve? Initially you'd be despondent, believing that your sentence amounted to life imprisonment. In time, you might grow hopeful of an early release, particularly if you know that an eighty-two-year-old was able to factor this number before the era of calculators and computers. If he was able to factor the number then you should also be able to do so. But maybe he knew more math than you do. Quite possibly, but basic arithmetic, along with an unknown amount of mental labor, is all you need to gain your freedom. You'd certainly get to know a few things about numbers, and you might discover that you had hidden talents.

My father often used this "prison question" during his lectures when he wanted to emphasize that something now known to us was at one time far from obvious, and that its attainment is still worthy of great admiration. By the way, here is the result of Landry's labors:

$$18{,}446{,}744{,}073{,}709{,}551{,}617 = 274{,}177 \times 67{,}280{,}421{,}310{,}721$$

RSA129

In the past, the intractability of factoring a number would have been viewed as somewhat disappointing because the prevailing spirit was one of complete optimism that all problems, no matter how hard they appeared to be, would eventually yield to analysis and ingenuity. But now, instead of being discouraged, we turn the tables around and exploit the very difficulty of factoring. This is what Ronald Rivest, Adi Shamir and Leonard Adleman, three students at the Massachusetts Institute of Technology (MIT), did in 1977 when they invented a famous cryptosystem based on the tremendous difficulty of factoring large numbers. That August, in a *Scientific American* article, Martin Gardner described this wonderful system and published a 129-digit number that came to be known as RSA129 (after the initials of the inventors' surnames and the length of the number). The reader was told that this number is the product of two primes. If one could factor the number one could decode a certain secret message and claim a $100 prize. (I thought this rather a paltry reward, but I could hear Dad saying,

"The real reward is the joy of solving the problem.") To quote Barry Cipra, writing in 1997,

> That $100 looked safe, at least for the next 20,000 years or so. . . . Not so. The RSA challenge yielded a mere 17 years later to a calculation that took less than a year from start to finish. A loosely organized group of factoring aficionados, numbering over 600 individuals in more than two dozen countries, nailed the 64- and 65-digit prime factors of RSA129 in an eight month effort, ending in April, 1994.

Although the 20,000-year time estimate proved way off the mark, this quote underlines the great difficulty of factoring numbers in excess of 100 digits. Even as larger and larger numbers come within the range of improved factoring techniques and faster machines, there will always be an infinity which lies beyond our reach.

The fact that the composite RSA129 was deliberately constructed as a product of two prime numbers is essential to the working of the cryptosystem, and something I hope you will find intriguing when I explain it later. When I first heard how the RSA system makes such a clever use of primes, I began to appreciate Dad's punning statement, "Prime numbers have become of prime importance."

In discussing Fermat's formula for generating primes, I mentioned two other mathematicians, Leonhard Euler, a great admirer of Fermat's work, and Father Mersenne. Both of these men also had something to say about methods for generating prime numbers. Euler hit upon the formula

$$E(n) = n^2 - n + 41$$

when searching for ways to produce primes. $E(n)$ is called the nth Euler number. For example, $E(1)$ is the first Euler number; its value is obtained by substituting 1 for n in the expression $n^2 - n + 41$.

So $E(1) = 41$ and the first Euler number is a prime. How about the second Euler number, $E(2)$? Well, $E(2) = 2^2 - 2 + 41 = 4 - 2 + 41 = 43$, so it is also a prime. Furthermore, $E(3) = 3^2 - 3 + 41 = 9 - 3 + 41 = 47$ is a prime, as is $E(4) = 53$. Incredibly, the first forty Euler numbers are all prime. However, you would be sadly mistaken if, on the basis of these first forty cases, you assumed that this remarkable formula generates primes for every natural number n. Alas,

$$E(41) = 1681 = 41 \times 41$$

is not prime. Neither is

$$E(42) = 1763 = 41 \times 43,$$

a product of twin primes. Euler's formula is a gallant attempt, but one which unfortunately does not generate prime numbers exclusively. Is the quest for a formula which generates *all* the primes a fruitless one?

Let us go back a century to Mersenne, who studied the numbers generated by the formula $M(n) = 2^n - 1$ when n is a natural number. It won't surprise you to learn that these numbers are known today as Mersenne numbers. Let's calculate the first nine of these numbers and see what observations we might make:

$$M(1) = 2^1 - 1 = 2 - 1 = 1$$
$$M(2) = 2^2 - 1 = 4 - 1 = 3$$
$$M(3) = 2^3 - 1 = 8 - 1 = 7$$
$$M(4) = 2^4 - 1 = 16 - 1 = 15 = 3 \times 5$$
$$M(5) = 2^5 - 1 = 32 - 1 = 31$$
$$M(6) = 2^6 - 1 = 64 - 1 = 63 = 3^2 \times 7$$
$$M(7) = 2^7 - 1 = 128 - 1 = 127$$
$$M(8) = 2^8 - 1 = 256 - 1 = 255 = 3 \times 5 \times 17$$
$$M(9) = 2^9 - 1 = 512 - 1 = 511 = 7 \times 73$$

Obviously Mersenne could not have asserted that his formula would always produce prime numbers. It might appear from this small amount of numerical evidence that if n is a composite natural number then $M(n)$ is also composite, an assertion we can render succinctly in the form n composite $\Rightarrow M(n)$ composite. (Recall that the symbol $\Rightarrow$ stands for "implies that.") This proposition is, in fact, true in general, and Mersenne would have known the little amount of algebra needed to prove it. I would love to show you how to do this, but I do not want to "blind you with science." Dad showed us in class how to prove this result, and emphasized what a marvelous thing it is to be able to prove a general result. He also said that it is good to look at a result from different angles. For example, you could say that this result says that if you substitute a composite value for n into the formula $2^n - 1$ you never get a prime number. Check that $M(10) = 2^{10} - 1 = 1023$ is indeed composite (see Appendix B, page 299). This is very easy. So there is no point substituting composite values for n if you want $M(n)$ to produce prime numbers. You must stick with prime values for *n;* it is necessary that n be prime for $M(n)$ to be prime. But is it always true that n prime $\Rightarrow M(n)$ prime?

In other words, is n being a prime sufficient to ensure that $M(n)$ is also a prime? Wouldn't it be great if this were true in general? The small amount of evidence we have from the above calculations shows us that when the prime values 2, 3, 5 and 7 are substituted for n into the formula for $M(n)$, the resulting Mersenne numbers are also prime. But as we know from our recent discussion of Euler's formula, the truth of a proposition for a finite number of cases does not prove the result in general. So let's experiment a little more. (I like this playful aspect of number theory that lets us explore by calculating.)

Let's substitute the next prime, 11, for *n,* and cross our fingers that this is a prime number: $M(11) = 2^{11} - 1 = 2047$. Now test it with the trial division algorithm. Since $\sqrt{2047} = 45.243\ldots$, we need to divide 2047, in turn, by the primes that are less than 45. If none of these 13 primes divides into 2047, then it is a prime, but if one of them does then it is a composite number. Unfortunately, 2047 is not a prime: I'll let *you* find its smallest prime factor. It will take some calculating, but finding the answer may help you appreciate how much time and effort the mathematicians of the sixteenth and seventeenth centuries expended on similar calculations. (You can verify your answer in Appendix B, page 300). They were rewarded by discovering some amazing truths about numbers. So the proposition n prime $\Rightarrow M(n)$ prime is not true in general. Although it is *necessary* for the number n to be prime for $M(n)$ to be also prime, sadly, as we have just seen, it is not *sufficient* to ensure that $M(n)$ is prime.

At one stroke a single "ugly" fact has destroyed a beautiful theory. This instance, where $2^{11} - 1$ is composite though 11 is a prime, is termed a counter-example to the general proposition. For a general proposition to be true, every single case must be true, which is a very stringent requirement; it takes only a single counter-example to make a proposition untrue in general. A generally true proposition is something to be admired.

If you substitute further prime numbers for n in Mersenne's formula, the resulting Mersenne numbers become very large and require much testing to see whether they are prime. Despite this, Mersenne asserted than $2^n - 1$ is prime for the prime values $n = 2$, 3, 5, 7, 13, 17, 19, 31, 67, 127 and 257, and for no other primes less than 257. He wasn't entirely right. The first eight of these values do give prime Mersenne numbers, but the 21-digit $M(67)$ is composite, a fact first discovered by the American mathematician Frank Cole. After two years of using his leisure time on Sundays, Cole factored $2^{67} - 1$. In 1903 he gave a completely silent "lecture" which consisted of two calculations. He first worked out $2^{67} - 1$ to be 147,573,952,589,676,412,927. Then he multiplied 193,707,721 by 761,838,257,287 and got the same result. Having shown that $M(67)$ is composite he then sat down to a standing ovation. Think of it—over two hundred days of calculations to factor a 21-digit number. When I asked *Mathematica* to write out the 21 decimal digits of $M(67)$, it did so in the blink of an eye, and when I asked it to factor this number it found the two prime factors in 1.792 seconds.

When I asked *Mathematica* if $M(127)$ is a prime, it returned the answer "True" in less than half a second. However, when I asked it to factor the 78-digit number $M(257)$, which is known to be a composite number, I got no immediate response other than an indication that the program was "running." It was still working away hours later. I eventually abandoned ship, but Dad put a faster computer to work on the same calculation beginning one Wednesday at one in the afternoon. The 78-digit number was still not factored almost a week later; at eight-thirty the following Tuesday morning the "running" message was still on the screen, indicating that the computer was still searching. The number $M(257)$ is known, since 1979, to be a product of three primes, the smallest of which is 535,006,138,814,359. Mersenne also missed some numbers; although the primes 61, 89 and 107 do not appear in his list, the numbers $M(61)$, $M(89)$ and $M(107)$ are, in fact, primes.

How was Mersenne able to make the claims he did in a time when, as far as we know, the art of factoring numbers was in its infancy? Near the end of the nineteenth century, the French mathematician Edouard Lucas found an efficient way of testing whether or not a Mersenne number is prime. This test, which was further refined by D. H. Lehmer in the twentieth century, is now known as the Lucas-Lehmer test. Because of its efficiency, searches to reveal a prime larger than the largest currently known are often confined to using this superfast test on Mersenne numbers.

For many people, the search for larger and larger prime numbers is something of a hobby. I can remember Dad saying in January 1998 that he had read on the Internet of how nineteen-year-old Ronald Clarkson, a college sophomore at California State University, became one of the many thousands of members of GIMPS (Great Internet Mersenne Prime Search). This group of amateurs uses their computers' "idle time" and software written (by mathematician George Woltman) specifically for the purpose of searching for large primes among the Mersenne numbers. Ronald was rewarded when after many days of toil his home computer found

$$M(3{,}021{,}377) = 2^{3{,}021{,}377} - 1$$

to be a prime number. This Mersenne number was the largest

prime known—until June 1999. Could you say how many digits are in its decimal representation? The answer is a staggering 909,526. If all the pages of the London *Times* were devoted to printing this prime number, there still would not be enough space. It is truly amazing that such a gigantic number has no factors other than itself and 1. On June 1, 1999, another GIMPS member, Nayan Hajratwala, a twenty-five-year-old technology consultant in Plymouth, Michigan, became the first person to find a megaprime, a prime number with a million or more digits. He found

$$M(6{,}972{,}593) = 2^{6{,}972{,}593} - 1$$

to be prime. This number has 2,098,960 digits—a megaprime indeed.

I have described three unsuccessful attempts to find a formula that generates prime numbers exclusively. Is there a formula that can do this? Apparently Jones, Sato, Wada and Wiens have discovered a complicated formula (if you must know, it's a polynomial of degree 25 in 26 variables) whose positive values are exactly all the prime numbers. I know nothing about the details, but it is absolutely fascinating that such a formula exists. I'm told that it can yield negative values also, such as −76, which is not a prime. But who cares, considering what it achieves?

How Many Prime Numbers Are There Less Than a Given Number?

This is an important question in its own right and from the point of view of cryptography. Many exact answers are known. There are exactly 25 primes less than 100, exactly 168 less than 1000 and exactly 1229 less than 10,000. There are 78,498 primes less than 1 million, 999,983 being the largest. As of January 2000, the best-known exact result is that there are exactly

> 2,220,819,602,560,918,840 primes less than
> 100,000,000,000,000,000,000.

However, in general, given a natural number *n*, it is not known exactly how many primes are less than or equal to *n*. If you study the random way in which the primes make their appearance as you proceed along the natural numbers, you may not find it very surprising that an exact formula still eludes mathematicians. What *is* surprising, however, is that one can estimate how many primes there are below each natural number. This famous result, known as the prime number theorem (PNT), was conjectured by mathematicians such as Karl Friedrich Gauss and Adrien Marie Legendre around 1800, based on their studies of prime number tables, but not proved until 1896.

This theorem estimates, for example, the number of primes less than 1 million at 72,382, which is not a bad approximation to the exact value of 78,498. Sometimes we can get by well enough with rough estimates. We can never hope to know population sizes, newspaper circulations or program ratings *exactly*, and are usually happy enough with estimates. We shall have occasion to use the PNT later, when we will have need to be convinced, for the purposes of discussing the RSA cryptosystem, that there is an abundance of primes of 100 digits or more.

6 The Arithmetic of Cryptography

If you have survived all the talk about prime numbers and didn't find the going too hard, you'll have no trouble reading this chapter and the next, which introduce you in an entertaining manner (I hope!) to the special arithmetic used in cryptography. This arithmetic is no harder to learn than the ordinary arithmetic of business that we're all familiar with; as you'll see, it's really quite straightforward and useful. This is the arithmetic I had to master when preparing my first project and I found it great fun once I got over its initial strangeness.

Don't worry if all the mathematical details are not jumping off the page at you—just skim to get the general picture. If, however, you've found the math a little too heavy and you haven't already skipped ahead to Chapter 8, you have my permission to go there now. There you'll find that the wonderful concepts of modern cryptography can be gleaned without in-depth knowledge of their mathematical underpinning.

This chapter has two aims. The first is to introduce one single piece of notation and terminology, which more than likely will be new to you. This is done by posing some puzzlelike problems that are simple and involve arithmetic calculations that will be familiar to most. I show you a little trick that will make your friends believe you have magicianlike mathematical capabilities. Interested?

Secondly, by describing the Caesar cipher in detail I reinforce the cryptographic language I have already used in the last chapter, and show you some of the nuts and bolts of enciphering and deciphering text. What I love about Caesar's system is that it's simple and that it allows so many of the main concepts and terminology of cryptography to be explained naturally. I hope you'll begin to see why I became so interested in the subject.

But first let me say a little about the central aim of cryptography and why it is used, although I know that most readers will already have a pretty good idea about the nature and aims of this science of secrecy.

Basically, cryptography is about keeping Nosy Parkers and/or more sinister characters with evil intentions from reading private messages intended for other people. Unfortunately, this is the way of the world, so great pains must be taken to turn private messages into gobbledygook before sending them to their destination. Then, should the aforementioned disreputable characters manage to intercept, or eavesdrop, or simply get a glimpse of the disguised message by some clever but illegitimate means, they will not immediately be able to make any sense of what they read. If you are knowledgeable enough you can encipher the message so that they will never be able to decipher it *in a realistic amount of time,* no matter how ingenious they may be and how unlimited their resources. That's about it in a nutshell.

In my first project report, I put it a little more formally:

> As long as there are creatures endowed with language there will be the desire for confidential communication—messages intended for a limited audience. Governments, companies and individuals have a need to send messages in such a way that only the intended recipient is able to read them. Generals send battle orders, banks wire fund transfers and individuals make purchases using credit cards.

I continued by saying that the central problem of cryptography is:

> How can a message be transmitted secretly to its intended recipient so that no unauthorized person obtains knowledge of its content?

Cryptography is the study of methods of sending messages in disguised form so that only those for whom they are intended can remove the mask and read their contents. In former times it was used mainly by the military and diplomatic services, but with the advent of electronic communication it is increasingly being used by financial institutions to ensure privacy in transactions between client and customer. One example of particular importance to e-commerce is the ability of buyers to make credit card purchases without vital details, such as the card's number, being compromised.

What Matters Is What's Left Over

Discussing the following puzzlelike problem allows us to introduce a simple notation which is used day in and day out by those who practice cryptography. After we go through a few examples to help you get used to the notation, you'll see how it is used to describe a system of enciphering and deciphering first used by Julius Caesar.

Which day of the week will it be 100 days from now if today is Sunday?

One way of solving this problem is to use a calculator to get that

$$\frac{100}{7} = 14.285714\ldots$$

Why divide 100 by 7? Because there are 7 days in a week. Since the result of our calculation is not an exact integer we conclude that 100 days contains 14 full weeks with a few days to spare. To calculate the number of days left over, or remaining, we work out that these 14 weeks will take $14 \times 7 = 98$ days to elapse. This done, we subtract the result of this calculation, 98, from the total number of days, 100, to get 2 as the number of days remaining. These two calculations can be combined:

$$100 - (14 \times 7) = 100 - 98 = 2$$

Equivalently,

$$100 = 14 \times 7 + 2$$

—a form which shows 100 as being made up of 14 multiples of 7 and a remainder of 2.

Perhaps the "short division"

$$\begin{array}{r} 7\underline{)100} \\ 14 + 2 \end{array}$$

taught in junior school reveals more quickly that $100 = 14 \times 7 + 2$.

In the language of weeks and days, $100 = 14 \times 7 + 2$ shows that 100 days consists of 14 full weeks and 2 extra days. So if today is Sunday then it will be a Tuesday in 100 days because after 14 full weeks it will be a Sunday again and the 2 extra days bring the day of the week to Tuesday.

If we ask similar questions, such as which day of the week it will be a thousand days from now or a million days from now, then we will need to do new calculations like the one above in order to answer them. But throughout we shall always be dividing by the number 7 because it measures the number of days in a week. When we are dealing with a positive divisor which remains constant from problem to problem, it is called the modulus ("mod" for short), a term that is a first cousin of a Latin word for "measure." If we are dealing with problems relating to clock times, the modulus will be either 12 or 24, while if we are dealing with problems involving angles the underlying modulus will be 360.

In the above calculation what really matters is the number 2 which is left over after 7 is subtracted 14 times from 100. It is this remainder 2 which points to the day of the week. In the notation of mathematics we can write

$$100 \bmod 7 = 2$$

to indicate that 2 is the remainder when 100 is divided by the fixed number, or modulus, 7.

Here is a charming story from Richard Guy, quoted in *Lure of the Integers* by Joe Roberts, which has some connection with our discussion. In Guy's words:

> There used to be an admission examination, the "11-plus," to British secondary schools. A question that was asked on one occasion was "Take 7 from 93 as many times as you can." One child answered, "I get 86 every time." I hope she got her place!

Did the examiners expect the students to know that, in our "mod" notation, 93 mod 7 = 2, or did they want them to say that you can take 7 from 93 thirteen times?

When working "with respect to" the modulus 7 there can be only seven possible remainders, namely 0, 1, 2, 3, 4, 5 or 6. It is useful to imagine these numbers written on a clock face in the way the numbers 1 to 12 ordinarily appear. Let the number 0 be at the top (where 12 is on the ordinary clock), and let the other six numerals be placed in order at equally spaced intervals around the perimeter of the clock. Call this clock a week-day clock, and imagine the different days of the week written under the seven numerals in small letters, with Sunday appearing below the 0 so that Saturday appears beneath the 6, which is to the left of the 0. If you wish you might also imagine that this clock has a single hand pivoted at the center of its face and pointing to the current day, only moving at the stroke of midnight.

If the hand of this clock points at the numeral 0 then today is Sunday. In 100 days' time the clock's hand will have made fourteen full rotations in the clockwise direction, and moved on a further two days, and so will point at the numeral 2. Hence 100 days from now it will be a Tuesday.

Let's use this clock to figure out which day of the week it was 100 days ago if today is Sunday. To go backwards in time, imagine the hand of the clock moving counterclockwise. Starting at 0, fourteen full rotations in the counterclockwise direction bring it back to 0. Continuing a further two days will bring it to the numeral 5, so if today is Sunday then 100 days ago it was a Friday.

Since we use 100 for the purpose of calculating which day it will be a hundred days into the future, we use -100 (minus 100) to calculate one hundred days into the past. If the "mod" operation is to indicate that 100 days ago it was a Friday, it must be that $-100 \bmod 7 = 5$. This allows the day of the week to be read directly.

You might argue, since $-100 = (-14) \times 7 + (-2)$, that $-100 \bmod 7$ should be -2, and you might reinforce your point by adding that -2 indicates the day of the week just as effectively as 5 does. The -2 says, "Go back two days from 0 (which represents Sunday) to arrive at Friday." You make a very good point, except that there are no negative numbers on the clock face.

The convention with the mod operation is that every number, be it a positive or a negative number or 0, when divided by the modulus leaves as a remainder either 0 or one of the positive numbers less than the modulus. In this way the answer always comes out as a number on the clock face.

From this viewpoint, -12 leaves a remainder of 2 when divided by 7. Why? Because $-12 = (-2) \times 7 + 2$. Thus $-12 \bmod 7 = 2$, a result which is easily understood by moving counterclockwise around the weekday clock. Note that $-2 \times 7 + 2$ can be interpreted as saying, "Starting at 0, make two full counterclockwise rotations to arrive at 0 again, and then go clockwise through 2 days."

Convince yourself that $-3 \bmod 7 = 4$ and $-13 \bmod 7 = 1$.

Here are a few more calculations involving the modulus 7:

$63 \bmod 7 = 0$	$8 \bmod 7 = 1$	$583 \bmod 7 = 2$
$647 \bmod 7 = 3$	$410 \bmod 7 = 4$	$691 \bmod 7 = 5$
$314 \bmod 7 = 6$	$4 \bmod 7 = 4$	$-8 \bmod 7 = 6$

Some explanations: $63 \bmod 7 = 0$ because $63 = 9 \times 7 + 0$, while $314 \bmod 7 = 6$ because $314 = 44 \times 7 + 6$. Furthermore, $4 \bmod 7 = 4$ because $4 = 0 \times 7 + 4$. This says that 7 goes into 4 zero times with a remainder of 4. In terms of the weekday clock, the hand has moved from 0, through 4 days, and now points at 4. Using the same clock, $-8 \bmod 7 = 6$ because starting from 0, the negative number -8 indicates a full counterclockwise rotation followed by a further rotation through one day in the same direction to arrive at 6. Alternatively, think of it as $-8 = -2 \times 7 + 6$ so that $-8 \bmod 7 = 6$.

When a number is divided by 7, it can leave one and only one of the remainders 0, 1, 2, 3, 4, 5 or 6. Should it leave the remainder 0 when divided by 7, as 63 does, then the number is divisible by 7. If the number leaves any of the other remainders when divided by 7, then that number is not divisible by 7.

Here are some more examples with respect to different moduli:

$$800 \bmod 360 = 80 \qquad 70 \bmod 11 = 4 \qquad 8678 \bmod 24 = 14$$

The first result could be interpreted as saying that an angle of 800° is the same as an angle of 80°, while the second result might tell us that if soccer teams are made up from 70 players then exactly 4 will be left on the sideline. The final result might inform us that if it is now four o'clock in the afternoon, then in 8678 hours it will be six o'clock in the morning.

A Party Trick

Here is a nice party trick perhaps best done with a calculator. It is based on a mathematical result called the *Chinese remainder theorem*. This result has many important (practical) applications besides the one you're about to see:

If someone tells me the three numbers remaining when she divides her age by 3, 5 and 7, respectively, then I can work out her age.

Here's how it is done. Suppose the person in question is aged a years. It is assumed that a is a natural number less than 105 (a restriction likely to be satisfied by most partygoers). The three remainders are a mod 3, a mod 5 and a mod 7. Suppose they work out to be x, y and z, respectively. Then, mysteriously (don't expect to understand why),

$$a = (70x + 21y + 15z) \bmod 105$$

This formula always works. Suppose, for example, that you, the reader, are aged seventeen. Then $a = 17$ with

$$\begin{aligned} a \bmod 3 &= 17 \bmod 3 = 2 \\ a \bmod 5 &= 17 \bmod 5 = 2 \\ a \bmod 7 &= 17 \bmod 7 = 3 \end{aligned}$$

Thus the three remainders which you give me are 2, 2 and 3, respectively. These are all I need to find your age. I set $x = 2$, $y = 2$ and $z = 3$ and, if necessary, with the aid of a calculator, I compute

$$\begin{aligned} 70x + 21y + 15z &= 70 \times 2 + 21 \times 2 + 15 \times 3 \\ &= 140 + 42 + 45 \\ &= 227 \end{aligned}$$

This done, I work out that $227 = 2 \times 105 + 17$ and so get $a = 227 \bmod 105 = 17$. Then I announce that you are seventeen years of age.

You can take it on faith that no other natural number less than 105 leaves this set of three remainders. Each natural number less than 105 has a unique triple of remainders which identify it. (Observe that $3 \times 5 \times 7 = 105$.)

If you really want to impress, train yourself to do all the arithmetic in your head. Some ages involve less mental calculation than others.

But can the remainders of numbers divided by 7 or any other modulus be worked out without first determining *how many times* the modulus goes into the number? Amazingly, yes. In some cases, even for colossal numbers, almost effortlessly. A special "arithmetic of remainders" (introduced by our old friend Karl

Friedrich Gauss when he was twenty-four years old) allows us to calculate remainders without having to do divisions. This is the arithmetic at the heart of cryptography, and I explain it in the next chapter. But the "mod" notation is all we need to describe some simple classical cryptosystems. Let's have a look.

The Caesar Cipher

One of the first military figures to encipher messages was Julius Caesar. He used the simple idea of substituting different letters of the alphabet for the actual letters of the message, or plaintext. This he did in a systematic fashion by shifting the alphabet forward three places so that a was enciphered as **d,** b as **e,** and so on, the three letters x, y and z being "wrapped around" and enciphered as **a, b** and **c,** respectively. This can be done efficiently by arranging the ciphertext alphabet beneath the plaintext alphabet as follows:

plaintext	a	b	c	d	e	f	g	...	v	w	x	y	z
ciphertext	**d**	**e**	**f**	**g**	**h**	**i**	**j**	...	**y**	**z**	**a**	**b**	**c**

Thus, without too much effort, he could encipher this message is top secret as **wklvphvvdjhlvwrsvhfuhw**. Here the blank spaces occurring in the ordinary plaintext have been suppressed. To better display the enciphering/deciphering relationship we write

this message is top secret → **wklvphvvdjhlvwrsvhfuhw**

The arrow indicates that the plaintext message transforms into the ciphertext cipher. The word "cipher" in this context means a secret message, whereas in Caesar's cipher it means a method of writing messages in secret.

The Caesar cipher is an example of a character or monographic substitution cipher since it enciphers by substituting single (mono) letters (characters) for other single letters.

To decipher the received cipher the recipient need only interchange the roles of the two alphabets, or if preferred shift back three characters, forming the following table:

ciphertext	**d**	**e**	**f**	**g**	**h**	**i**	**j**	...	**y**	**z**	**a**	**b**	**c**
plaintext	a	b	c	d	e	f	g	...	v	w	x	y	z

Then

wklvlvkrzzhghflskhu → thisishowwedecipher

which is easily seen to be "this is how we decipher." We could write

thisishowwedecipher ← **wklvlvkrzzhghflskhu**

and so keep the ciphertext on the right. Here the arrow pointing from right to left indicates that decipherment, as opposed to encipherment, has taken place.

Decipher **wubdjdlq** assuming it has been enciphered using a 3-shift as just described. (Verify your answer in Appendix B, page 300.)

Occasionally a word becomes another recognizable English word when it is enciphered. A good example of this is the word "cold." Encipher it and see. (Once again, check Appendix B, page 300, to see if you got it right.)

Clearly one can use shifts other than 3. Hal, the name of the computer in Arthur C. Clarke's *2001: A Space Odyssey,* is obtained from IBM by a forward shift of 25—or, equivalently and much more conveniently, by a backward shift of 1. Of course, the longer a word, the less likely it is to be enciphered as another English word. One of the longest of such words in English is "abjurer." In a 13-shift cipher it becomes "nowhere." You might like to determine what happens to "pecan" in a 4-shift cipher or to "sleep" in a 9-shift cipher. (And you might like to check your answer in Appendix B, page 300.)

There is no need to have two separate tables like the ones I used for illustrative purposes. One will do. To encipher, read the first table from top to bottom; to decipher, read it from bottom to top. An attractive alternative is to make a Caesar wheel from two cardboard discs, one slightly larger than the other (as I did for my first project). Place the smaller one on top of the larger one with their centers superimposed, and mark off 26 even (very even!) sectors by lines extending from the center to the edge of the larger disc. Mark the letters of the alphabet (in order) in these sectors near the edges of each disc. Pin the discs together at their centers so that they are free to spin independently. It is now easy to set up any of the 26 possible letter shifts. For a 5-shift simply turn the inner disc or wheel so that its *f* is beneath the *a* on the outer disc. All the other letters will be aligned correctly.

The Caesar cipher is a cryptosystem that can be thought of as applying solely to alphabetic characters. It can be implemented using tables or wheels, or even on a computer, without ever having to use numbers. However, it is best, for at least two reasons, if at their core cryptosystems use numbers. The first reason is that computers are ideally suited to handling numbers, even huge ones, with great speed. The second is that there is a large body of mathematical functions available that can transform numbers into other numbers in complicated ways. This makes it possible to design secure but efficient systems. No doubt very secure systems can be designed which make no use of numbers, but can they function efficiently? There is no point in having the instruction "sell shares at 140" cloaked in almost impenetrable secrecy if the share price has dropped through the floor by the time the message is deciphered.

Numbers for Letters

The table

a	b	c	d	e	f	g	h	i	...	v	w	x	y	z
0	1	2	3	4	5	6	7	8	...	21	22	23	24	25

assigns to each of the twenty-six (lower-case) letters of the alphabet a unique whole number between 0 and 25. These numbers are called the *numerical equivalents* of the letters. The letter **e** has the numerical equivalent 4. You'll see why the letters of the alphabet are assigned the numbers 0 to 25 as opposed to 1 to 26 very soon. It will also become clear that the letters in the top row of the table are sometimes used as plaintext characters and sometimes as ciphertext characters.

Thinking in terms of the numbers instead of the letters, what does the Caesar cipher described earlier do? Add 3 to each number? True most of the time, but not always because of the "wraparound" at the end of the alphabet. Adding 3 to the number 4 gives the number 7. Since 4 is the numerical equivalent of the letter **e**, and 7 is the numerical equivalent of the letter **h**, **e** gets enciphered as **h**, which is correct. However, simply adding 3 to 24, which is the numerical equivalent of **y**, will not encipher this letter correctly as 27 is not a numerical equivalent of any character.

So how to iron out this wrinkle? What to do with those numbers that exceed 25? Since **y** is supposed to encipher to **b**, which has numerical equivalent 1, the number 27 must be converted to a 1. Can you see how to turn 27 into 1?

Remember the "mod" notation? Isn't 27 mod 26 = 1? Numbers that exceed 25 when "reduced by the size of the alphabet" (which is 26 in this case) will become the correct numerical equivalents of the corresponding ciphertext characters. The number 26 becomes 0, while 27 becomes 1, and 28 becomes 2. If 27 is regarded as 1, then the letter **y** is enciphered correctly as **b.**

Instead of saying "add 3 to each numerical equivalent," the instruction is "add 3 to each numerical equivalent, reducing by 26 any number larger than 25."

The arithmetic, or system of numerical calculation, that does this correctly is arithmetic with respect to the modulus 26. The phrase "with respect to the modulus" is usually abbreviated "modulo" so, with this new terminology, the correct arithmetic is *arithmetic modulo* 26. When working modulo 26, the only possible remainders are numbers in the range 0 to 25. It is for this reason that the letters of the alphabet are assigned the numbers 0 to 25 rather than 1 to 26 or some other range. To do this arithmetic in your head just think of a 26-hour clock with the numbers 0 to 25 arranged in the usual clocklike manner. Then it is easy to see why

$$23 + 3 = 0,\ 24 + 3 = 1,\ 25 + 3 = 2$$

However, it is sometimes convenient to allow numbers, even negative ones, outside the range 0 to 25 to appear in calculations when working modulo 26, as their use often makes things simpler. You'll see some examples later. If you need to, you can always use a number between 0 and 25 to perform the same function.

The Enciphering and Deciphering Formulae

To describe concisely the procedure for the Caesar cipher with a shift of 3, it helps to let P stand for the numerical equivalent of a typical plaintext character, and C stand for the numerical equivalent of the corresponding ciphertext character. Then the equation

$$C = (P + 3) \bmod 26$$

gives the general encipherment rule. Isn't this a simple and a practical use of the mod notation?

It is important to bear in mind that though P and C are written as letters, they are, in fact, numbers. The P-number of the letter **x** is 23, so the corresponding C-number is

$$C = (23 + 3) \bmod 26 = 26 \bmod 26 = 0$$

Since 0 is the numerical equivalent of **a,** this scheme gives the correct encipherment **x** → **a.**

When I eventually knuckled down to programming, one of my first tasks was to get this end of the Caesar system working. It wasn't that hard to do, but I had to learn how to split a piece of text into single characters, assign each character its numerical equivalent, find the corresponding ciphertext numerical equivalents (the C-numbers) and then turn all these into single ciphertext characters. This last step can be omitted by simply leaving the plaintext enciphered as a string of numbers, but I preferred to show English text being enciphered into meaningless text. It was great fun learning the *Mathematica* commands to do all this, arranging them into a sequence of instructions and seeing chunks of ordinary English being transformed into an unreadable mess. I know now that a lot of the programming I did was fairly primitive, but just seeing it work was a real kick. To be sure that I was enciphering properly I had to convert the chaos of seemingly random ciphertext into the "order" it once was.

So how to decipher with a Caesar system which uses a forward shift of 3? Clearly, *shift back* 3, taking care to wrap around when the need arises. So the deciphering rule is described by

$$P = (C - 3) \bmod 26$$

To decipher the letter **b,** which has a C-number of 1, calculate

$$P = (1 - 3) \bmod 26 = -2 \bmod 26 = 24$$

Since 24 is the numerical equivalent of **y,** the ciphertext character **b** deciphers as the plaintext character **y,** as it should.

Why is $-2 \bmod 26 = 24$? On a clock face displaying the numbers 0 to 25 in the usual order, moving two steps counterclockwise from the number 0 brings us to the number 24. To confirm this result, think of a (backward) shift of -3 as a forward shift of 23. Then

$$P = (1 + 23) \bmod 26 = 24 \bmod 26 = 24$$

as before.

Of course, Caesar didn't have to confine himself to a shift of 3. Because the lower-case alphabet has 26 letters he could have used any of the 26 different possible shifts of the alphabet, though a shift of 0 doesn't do any enciphering at all. If the letter s is used to stand for the size of a typical shift, then the formula

$$C = (P + s) \bmod 26$$

describes the enciphering rule for a Caesar cipher with forward shift s. In the original example above, $s = 3$.

It should be clear when s is itself and when it is acting as a number. For example, for a Caesar cipher with a shift of 10 the letter **s** gets enciphered as **c,** as is easy to see. The numerical equivalent of the letter **s** is $P = 18$, while the size of the shift in the formula is $s = 10$. Thus

$$C = (18 + 10) \bmod 26 = 28 \bmod 26 = 2$$

The C-number 2 is the numerical equivalent of the letter **c**.

Clearly, the formula

$$P = (C - s) \bmod 26$$

describes the deciphering rule corresponding to the enciphering rule for a general Caesar cipher with a forward shift of size s.

Keys

In the nineteenth century Auguste Kerckhoffs (1835–1903) enunciated **Kerckhoffs's principle:** The security of a cryptosystem must not depend on keeping the cryptographic algorithm secret. The security depends only on keeping both keys secret.

For the general Caesar cryptosystem described earlier, the number s representing the size of the forward shift employed is termed the enciphering key. If K_E is used as a shorthand for "enciphering key," then for the general Caesar cipher,

$$K_E = s$$

The corresponding deciphering key, representing the size of the backward shift required to achieve deciphering, is the number $-s$. If K_D denotes "deciphering key," then for the general Caesar cipher,

$$K_D = -s$$

Between them, K_E and K_D are termed the keys of a cryptosystem. The Caesar cipher based on a forward shift of 3 has keys $K_E = 3$ and $K_D = -3$.

In the context of the Caesar system, Kerckhoffs's principle states that the security of the system should not depend on Caesar's enemies' ignorance of his general method of secret writing. It is to be expected that eventually his system will become known to people other than those originally in the know. From the very outset it is known to the generals with whom he is using it for private communication, and there is always the possibility that one of them will become an enemy at some stage. The security of the system must depend on keeping both the enciphering and deciphering keys secret.

Cryptanalysis

In practice, a Caesar cipher could never be considered a secure cryptosystem. Keeping its keys secret is not enough. If it is known that it is being used, then it can be broken, quite easily by analyzing captured ciphertext. One approach checks all possible values of $-s$ until a decipherment results in readable plaintext. This form of attack searches through the set of all possible keys—what is called the keyspace—until the correct one is found. In this case it is feasible even by hand because the size of the keyspace is only 25.

Another approach to breaking this system is to use frequency analysis based on the table, which gives the percentage of occurrence of each of the letters of the alphabet in typical English plaintext.

Character	a	b	c	d	e	f	g	h	i	j	k
Relative frequency	7	1	3	4	13	3	2	3	8	<1	<1

l	m	n	o	p	q	r	s	t	u	v	w	x	y	z
4	3	8	7	3	<1	8	6	9	3	1	1	<1	2	<1

Calculate the frequencies of each of the ciphertext characters. If there is enough ciphertext available, then the ciphertext character with the highest frequency will normally correspond to the plaintext letter e, as this letter occurs the most frequently in English plaintext. The shift size s is then determined by calculating the number of steps necessary to transform e into the corresponding ciphertext character.

The science of breaking cryptosystems is known as *cryptanalysis*.

Keeping Your Distance

By this stage you may be thinking that using numbers to represent letters is making everything more long-winded and creating an awful lot of complication where there seems to be no need for any. However, the numbers approach pays off when plaintext is enciphered in a more sophisticated way, to put more distance between a legitimate user and the ever-present enemy.

As a first step in designing a more complex cryptosystem, one might use a much larger "alphabet," including both the ordinary lower- and upper-case characters, along with the usual punctuation symbols. Often the alphabets used are of size 128 or 256 symbols. If n is the number of symbols used in the alphabet, then the enciphering and deciphering transformations for a cryptosystem based on a Caesar cipher with shift s are given by

$$C = (P + s) \bmod n \; ; \; P = (C - s) \bmod n$$

respectively. Now all arithmetic is done modulo n rather than modulo 26.

Even though the keyspace is much larger, this generalized Caesar system is still vulnerable to a computer search for s. A frequency analysis attack, on the other hand, requires much more preliminary statistical work to establish the relative frequencies of the occurrence of symbols from the larger plaintext alphabet.

What else can be done besides increasing the alphabet size? Why not try to improve on the Caesar system? This system essentially adds a fixed number s to each P-number and reduces the answer modulo n to get a natural number less than the modulus. How about multiplying instead of adding?

Let us experiment on the lower-case alphabet, so we are back to working modulo 26. In practice, the general modulus n would be much larger. Let's encrypt a typical P-number by multiplying it by, say, 5. Then the enciphering rule says that

$$C = 5P \bmod 26$$

which is very simple.

The following array shows what happens to the twenty-six P-numbers 0 to 25 when each is enciphered according to this rule:

P	$\rightarrow$	C	P	$\rightarrow$	C
0	$\rightarrow$	0	13	$\rightarrow$	13
1	$\rightarrow$	5	14	$\rightarrow$	18
2	$\rightarrow$	10	15	$\rightarrow$	23
3	$\rightarrow$	15	16	$\rightarrow$	2
4	$\rightarrow$	20	17	$\rightarrow$	7
5	$\rightarrow$	25	18	$\rightarrow$	12
6	$\rightarrow$	4	19	$\rightarrow$	17
7	$\rightarrow$	9	20	$\rightarrow$	22
8	$\rightarrow$	14	21	$\rightarrow$	1
9	$\rightarrow$	19	22	$\rightarrow$	6
10	$\rightarrow$	24	23	$\rightarrow$	11
11	$\rightarrow$	3	24	$\rightarrow$	16
12	$\rightarrow$	8	25	$\rightarrow$	21

Before we begin enthusing about the two wonderful columns of C-numbers, let's make sure we understand how these numbers appear. For example, the P-number 22 becomes the C-number 6 because

$$5 \times 22 = 110 = 4 \times 26 + 6 \qquad \text{so} \qquad (5 \times 22) \bmod 26 = 6$$

Verify a few more; well, at least one more. The array shows that the set

$$\{0, 1, 2, 3, 4, 5, 6, 7, 8, 9, 10, 11, 12,$$
$$13, 14, 15, 16, 17, 18, 19, 20, 21, 22, 23, 24, 25\}$$

is transformed by the enciphering rule $C = 5P \bmod 26$ into the set

$$\{0, 5, 10, 15, 20, 25, 4, 9, 14, 19, 24,$$
$$3, 8, 13, 18, 23, 2, 7, 12, 17, 22, 1, 6, 11, 16, 21\}$$

What's great about this is that all original twenty-six P-numbers reappear as C-numbers. This is significant because each P-number has a unique C-number—as it should—and, just as importantly, each C-number corresponds to a unique P-number. This kind of correspondence between the set of P-numbers and the set of C-numbers is very aptly described as a "one-to-one" correspondence. (Later, we'll see that matters work out somewhat more disappointingly when we multiply by 4 instead of 5.) Whereas the Caesar encipherment rule rotated the twenty-six P-numbers, the current encipherment rule permutes these twenty-six P-numbers. Notice how nicely this enciphering process jumbles the numbers.

If the numbers in the above array are replaced by their corresponding letters, the plaintext alphabet

{a, b, c, d, e, f, g, h, i, j, k, l, m, n, o, p, q, r, s, t, u, v, w, x, y, z}

is seen to encipher into the ciphertext alphabet

{a, f, k, p, u, z, e, j, o, t, y, d, i, n, s, x, c, h, m, r, w, b, g, l, q, v}

Thus the following enciphering/deciphering table is obtained:

plaintext	a	b	c	d	e	f	g	h	i	...	v	w	x	y	z
ciphertext	**a**	**f**	**k**	**p**	**u**	**z**	**e**	**j**	**o**	...	**b**	**g**	**l**	**q**	**v**

This table, consisting of characters alone, could be used to carry out all the enciphering and deciphering without ever having to make any use of numbers. An example of encipherment is

evacuate immediately → **ubakwaruoiiupoarudq**

while an example of decipherment is

buy oil shares ← **vekimxoratgo**

Constant looking up of a table can be very time-consuming, particularly when much larger alphabets are involved. Although it may not appear so yet, it is preferable to work as much as possible with numbers rather than characters. This being so, an interesting question now arises. What is the deciphering rule corresponding to the enciphering rule $C = 5P \bmod 26$? How do we go from a C-number, such as 22, back to the P-number that generated it? Consulting the above array of numbers and reading from right to left to find that the corresponding P-number is 20 is not considered to be legitimate because it involves "table lookup." What we need is a mathematical rule which, given a C-number, produces the corresponding P-number.

If it were ordinary arithmetic the solution would be to undo the forward "multiply by 5" by doing the opposite, which is "divide by 5." However, the fact that numbers also get reduced modulo 26 in the forward process proves something of an obstacle to this plan. For example, dividing the C-number 22 by 5 produces 4.4, which is not in the set of P-numbers. Decimal numbers don't normally appear in cryptographic calculations, nor do even the simplest of fractions, such as $1/3$. It's just the natural numbers most of the time, with the occasional negative whole number allowed for convenience.

So how to undo the forward action of "multiply by 5 modulo 26"? I'm going to tell you how it is done, and the method will appear very mysterious. I'm only going to hint at an explanation of why it works. It is not that it is hard to explain, but to do so properly would take us too far from our main theme.

To convert the *C*-numbers back to their original *P*-counterparts, multiply them by 21 modulo 26.

YOU: Did you say multiply? Surely there must be a division involved somewhere. And where does this magical number 21 come from?

ME: Yes, I did say multiply and yes, you are right—there is a division involved somewhere.

YOU: There doesn't appear to be.

ME: Agreed, but believe me that multiplying by 21 modulo 26 is equivalent in this arithmetic to multiplying by the fraction $1/5$ in ordinary arithmetic.

YOU: I see. Well, I don't see, but I do understand that multiplying by $1/5$ in normal arithmetic is the same as dividing by 5.

ME: Yes, in everyday arithmetic the number $1/5$ is called the multiplicative inverse of 5.

YOU: I remember. The number $1/5$ is the multiplicative inverse of 5 because when 5 is multiplied by $1/5$, the number 1 is obtained.

ME: Exactly. Now in arithmetic modulo 26, the number 21 is the multiplicative inverse of 5 because when 5 is multiplied by 21 and the result reduced modulo 26 the number 1 appears.

YOU: That's kind of fun because you can't get the inverse by just "turning the number upside down," as in ordinary arithmetic. Let me verify what you say. In ordinary arithmetic, $5 \times 21 = 105$, and since $105 = 4 \times 26 + 1$, it follows that $(5 \times 21) \bmod 26 = 1$.

ME: Impressed?

YOU: Yes, but I'd be more so if you could explain how you knew that 21 is the multiplicative inverse of 5 modulo 26, rather than producing it like a magician's white rabbit.

ME: I'm not going to tell you all my secrets, but you are right when you say that 21 is the multiplicative inverse of 5 modulo 26 because 21 is the only natural number between 0 and 25 that has this property.

YOU: Thank you, I didn't realize I was being profound. I'm afraid I automatically assumed that the number 5 would have only one inverse.

ME: If you allow natural numbers greater than the modulus 26 and/or negative whole numbers, then it has an infinite number of "inverses," two of which are 25 and 47, but this is beside the point.

YOU: If you say so. I'm going to check that your deciphering rule really works. I'll try it on the C-number 22.

The following calculation:

$$\begin{aligned} 22 \times 21 &= 462 \\ &= 17 \times 26 + 20 \\ \Rightarrow (22 \times 21) \bmod 26 &= 20 \end{aligned}$$

verifies that the C-number 22 corresponds to the P-number 20, as it should. Great!

The deciphering rule giving the P-number corresponding to a C-number enciphered by $C = 5P \bmod 26$ is

$$P = 21C \bmod 26$$

Thus $K_D = 21$ is the deciphering key.

Summary: This particular cryptosystem enciphers via

$$C = 5P \bmod 26$$

and deciphers via

$$P = 21C \bmod 26$$

Its keys are

$$K_E = 5 \; ; K_D = 21$$

These keys are called single parameter keys because each is specified by a single number.

Let us see what happens when we try enciphering using the formula

$$C = 4P \bmod 26$$

The following array shows the results of enciphering the numbers 0 to 25 according to this rule:

P	$\rightarrow$	C	P	$\rightarrow$	C
0	$\rightarrow$	0	13	$\rightarrow$	0
1	$\rightarrow$	4	14	$\rightarrow$	4
2	$\rightarrow$	8	15	$\rightarrow$	8
3	$\rightarrow$	12	16	$\rightarrow$	12
4	$\rightarrow$	16	17	$\rightarrow$	16
5	$\rightarrow$	20	18	$\rightarrow$	20
6	$\rightarrow$	24	19	$\rightarrow$	24
7	$\rightarrow$	2	20	$\rightarrow$	2
8	$\rightarrow$	6	21	$\rightarrow$	6
9	$\rightarrow$	10	22	$\rightarrow$	10
10	$\rightarrow$	14	23	$\rightarrow$	14
11	$\rightarrow$	18	24	$\rightarrow$	18
12	$\rightarrow$	22	25	$\rightarrow$	22

Notice anything about the two C-number columns? They are identical. For example, the P-number 9 becomes the C-number 10 because

$$4 \times 9 = 36 = 1 \times 26 + 10 \Rightarrow (4 \times 9) \bmod 26 = 10$$

while the P-number 22 also becomes the C-number 10 because

$$4 \times 22 = 88 = 3 \times 26 + 10 \Rightarrow (4 \times 22) \bmod 26 = 10$$

This is not good. The table shows that the ordered set

$$\{0, 1, 2, 3, 4, 5, 6, 7, 8, 9, 10, 11, 12, 13, 14, 15, 16, 17, 18, 19, 20, 21, 22, 23, 24, 25\}$$

gets transformed by this enciphering rule into the ordered set

$$\{0, 4, 8, 12, 16, 20, 24, 2, 6, 10, 14, 18, 22\}$$

twice over. The set of *C*-numbers is not the original set of *P*-numbers rearranged in some order, but just one "half" of that set. This time the correspondence between the set of *P*-numbers and the set of *C*-numbers is not a one-to-one correspondence but a "two-to-one" correspondence, since two *P*-numbers correspond to every *C*-number. As verified, the numbers 9 and 22 both correspond to 10, which in terms of characters, means that the letters j and w both encipher as **k**. Such an enciphering rule is a disaster because the ciphertext character **k,** as is the case with each of the others, cannot be deciphered uniquely. Clearly this won't do.

Things are much worse when every *P*-number is multiplied by 13 modulo 26. Do a few multiplications and you'll see why.

I hope you are curious as to how 21 is found to be thc (multiplicative) inverse of 5. One method is to multiply each of the numbers between 1 and 25 by 5 modulo 26 to see which produces an answer of 1. There is no point in your doing this for 5, now that you know the answer, so try 7 instead.

When you have found the inverse of 7 modulo 26, try to find the inverse of 4 modulo 26. (You need do no calculations as they are all done in the last array.) You'll get a surprise.

So now we have stumbled upon something very interesting by just experimenting. When we multiply *P*-numbers by 5 modulo 26 we get a one-to-one correspondence between the set of *P*-numbers and the set of *C*-numbers, and so obtain a valid cryptosystem, even if it is only a very simple one. On the other hand, when we multiply by 4 modulo 26 we fail to get a one-to-one correspondence that is so necessary to be able to encipher and decipher uniquely.

We can say that the enciphering rule

$$C = mP \bmod 26$$

works when $m = 5$ but not when $m = 4$, and we regard the number 5 as a "good" m-value and the number 4 as a "bad" m-value.

If we were to do some more playful experimenting, just trying different m values, then we'd find that the thirteen numbers 2, 4, 6, 8, 10, 12, 13, 14, 16, 18, 20, 22 and 24 are bad m-values, while the twelve numbers 1, 3, 5, 7, 9, 11, 15, 17, 19, 21, 23 and 25 are good m-values. For the modulus 26 there are only twelve good m-values. Granted, using the number 1 as a multiplier achieves no scrambling and so no enciphering, but it is still considered a good m-value because it gives a one-to-one correspondence between the set of P-numbers and the set of C-numbers (the identical correspondence).

So what property must a multiplier m have in order to guarantee a one-to-one correspondence between the set of P-numbers and the set of C-numbers? The answer is not so obvious, but it is simple. The number m must have no factor other than 1 in common with the modulus 26.

If this condition is satisfied, things work; if not, they don't. It is a necessary *and* sufficient condition. Check this condition on the numbers above. All the numbers 2, 4, 6, 8, 10, 12, 13, 14, 16, 18, 20, 22 and 24 have either the factor 2 or 13 in common with 26, while the numbers 1, 3, 5, 7, 9, 11, 15, 17, 19, 21, 23 and 25 have no factor in common with 26 other than the trivial factor 1.

Two numbers are said to be *relatively prime* to each other whenever they have no nontrivial factor in common. (The answer to the two jars puzzle gives another insight into this important concept.) Hence each of the numbers 1, 3, 5, 7, 9, 11, 15, 17, 19, 21, 23 and 25 is relatively prime to the number 26. In this terminology, the condition for a multiplier m to guarantee a one-to-one correspondence between the set of P-numbers and the set of C-numbers is that m be relatively prime to the modulus 26.

Thus there are only twelve choices for the multiplier m in the encipherment rule

$$C = mP \bmod 26$$

What about choosing a number m larger than 26? Try $m = 35$, and verify that it gives the same encipherment as $m = 9$. There is nothing new to be gained from using numbers bigger than the modulus. There really are only twelve good m-values. That said, how can we make things a tad more complicated while still working with the twenty-six characters of the lower-case alphabet?

Why not combine a multiplication with a shift? For example, we could follow multiplication by m with an addition (or shift) of s, so that the general encipherment rule is

$$C = (mP + s) \bmod 26$$

The enciphering key K_E associated with this enciphering rule is specified by two numbers: the multiplier m, which must be relatively prime to 26, and the shift s, which can be any number between 0 and 25. Thus $K_E = \langle m, s \rangle$, read as "the enciphering key m comma s," is a two-parameter key associated with the encipherment rule which gives a one-to-one correspondence between plaintext and ciphertext. When $m = 1$, the encipherment rule reverts to a Caesar cipher with shift s. Since there are twelve choices for the first parameter m (the twelve "good" m-values) and twenty-six for the second parameter s, there are $12 \times 26 = 312$ different keys in the keyspace.

When $K_E = \langle 5{,}18 \rangle$ (read as "the enciphering key 5 comma 18"), the encipherment rule is

$$C = (5P + 18) \bmod 26$$

It is straightforward to check that the set of P-numbers

$$\{0, 1, 2, 3, 4, 5, 6, 7, 8, 9, 10, 11, 12, 13,\\ 14, 15, 16, 17, 18, 19, 20, 21, 22, 23, 24, 25\}$$

is transformed by this enciphering rule into the ordered set

$$\{18, 23, 2, 7, 12, 17, 22, 1, 6, 11, 16, 21,\\ 0, 5, 10, 15, 20, 25, 4, 9, 14, 19, 24, 3, 8, 13\}$$

of C-numbers. Notice that the correspondence is one-to-one in nature.

Now for the interesting part. What is the deciphering key K_D corresponding to $K_E = \langle 5,18 \rangle$? Answer:

$$K_D = \langle 21,12 \rangle$$

(read as "the deciphering key 21 comma 12"). Thus the deciphering rule is

$$P = (21C + 12) \bmod 26$$

I hear you protest, "Another white rabbit out of the magician's hat!" Yes, but you have seen the rabbit number 21 before, undoing the forward action of multiplying by 5 modulo 26. For the moment I'm going to keep to myself how the number 12 gets in on the deciphering act. Here is how the deciphering works on the C-number 13:

$$P = (21 \times 13 + 12) \bmod 26 = 285 \bmod 26 = 25$$

since $285 = 10 \times 26 + 25$. If you check, you'll find that the P-number 25 does encipher as 13 under $C = (5P + 18) \bmod 26$.

Summary: For the system just described, the enciphering rule is

$$C = (5P + 18) \bmod 26$$

while the corresponding deciphering rule is

$$P = (21C + 12) \bmod 26$$

and the keys are $K_E = \langle 5,18 \rangle$ and $K_D = \langle 21,12 \rangle$.

Grouping Characters

Up to now we have been breaking plaintext into single characters, assigning each a numerical equivalent, or P-number, and enciphering that number, via some mathematical rule, to obtain a corresponding C-number. When we do this we can regard the plaintext as having been enciphered (as numbers). If we wish, we can assign these C-numbers to their corresponding characters to obtain an encryption of the plaintext into "true" ciphertext.

It should be clear that we can put much more distance between us and our enemy if we break plaintext into blocks of letters rather than into single characters. For example, working with the 26-character lower-case alphabet (for illustrative purposes only), we could break the plaintext

go placidly amid the noise and haste and
remember what peace there may be in silence

which contains sixty-nine characters (ignoring the blank spaces) into blocks of two:

go pl ac id ly am id th en oi se an dh as te an dr
em em be rw ha tp ea ce th er em ay be in si le nc ex

The x is appended to the last character e so that it, too, belongs to a complete block. In practice, there is no need for this because a blank character is part of the underlying alphabet. Blocks consisting of two characters are known as digraphs.

How is each digraph enciphered into a unique "ciphertext unit"? There are 26 × 26 = 676 possible digraphs, beginning with aa, ab, ac, . . . , ba, bb, . . . and ending with . . . ya, yb, . . . , zx, zy, zz. Why not assign each of them a unique number between 0 and 675 inclusive, and encrypt these numbers along lines similar to those for single characters?

A simple but effective scheme of assigning a unique number to each digraph is to assign the digraph aa the number 0, the digraph ab the number 1, and so on, with the digraph zz assigned the number 675. There is no need to look up a table to find the numerical equivalent (P-number) associated with a digraph: to determine the P-number for the digraph go, refer to the original table of numbers (associated with the 26 lower-case characters) to find the numerical equivalent of g as 6, and the numerical equivalent of o as 14. Then the P-number assigned to go is

$$P = (6 \times 26) + 14 = 170$$

Simple. Can you explain why this is correct? (To confirm your explanation, see Appendix B, page 301).

The P-number assigned to og is

$$P = (14 \times 26) + 6 = 370$$

With this rule for assigning numerical equivalents, a program need only know the numbers associated with each of the single characters in the digraph, take note of their order and compute the P-number as is done above.

The P-numbers of the set of digraphs given above are as follows:

170 401 002 211 310 012 211 501 117 372
472 013 085 018 498 013 095

116 116 030 464 182 509 104 056 501 121
116 024 030 221 476 290 340 127

There's Safety in (BIG) Numbers

So how might we encipher these P-numbers? Well, we could use a "multiply and shift" encipherment rule like

$$C = (mP + s) \bmod n$$

as we did for single characters. But

1. What is the correct modulus n?
2. What multiplier m should be chosen?
3. What shift constant s can be used?

Answers:

1. The modulus n is 676 because the 676 distinct digraphs are assigned the whole numbers from 0 to 675.
2. The multiplier m may be any number between 1 and 675, provided it is relatively prime to 676. (There are 312 such numbers.)
3. The shift constant s may be any whole number in the range 0 to 675.

The keyspace for this cryptosystem has size $312 \times 676 = 210{,}912$, which is considerably larger than the sizes of any of the previous keyspaces.

For example, choosing the key

$$K_E = \langle 159{,}580 \rangle$$

makes the enciphering rule

$$C = (159P + 580) \bmod 676$$

Then the C-number corresponding to the P-number 170 (which is the numerical equivalent of the digraph **go**) is calculated as follows:

$$\begin{aligned} C &= (159 \times 170 + 580) \bmod 676 \\ &= 27{,}610 \bmod 676 \\ \Rightarrow C &= 570 \end{aligned}$$

since 27,610 = 40 × 676 + 570.

How is 570 expressed as "true" ciphertext? Because

$$570 = (21 \times 26) + 24$$

and 21 and 24 are the numerical equivalents of the letters **v** and **y**, respectively, the ciphertext unit corresponding to "go" is **vy**. Thus

go → vy

Using the encipherment rule the C-numbers corresponding to the plaintext P-numbers are

570 119 222 329 522 460 329 471 255 240
592 619 575 062 670 619 137

096 096 618 672 450 391 216 020 471 215
096 340 618 567 552 046 560 493

This jumble of numbers can be converted into a string of ciphertext digraphs to give

vy ep io mr uc rs mr sd jv jg wu xv wd ck zu xv fh
ds ds xu zw ri pb ii au sd ih ds nc xu vv vg bu vo sz

Thus the encipherment of the plaintext

go placidly amid the noise and haste and
remember what peace there may be in silence

is

vyepiomrucrsmrsdjvjgwuxvwdckzuxvfh
dsdsxuzwripbiiausdihdsncxuvvvgbuvosz

Here, without a word of explanation (but see Appendix C, page 305), is the deciphering rule:

$$P = (659C + 396) \bmod 676$$

This means that the deciphering key is $K_D = \langle 659,396 \rangle$. Let's decipher the C-number 493, which is the final one above, using this rule:

$$\begin{aligned} P &= (659 \times 493 + 396) \bmod 676 \\ &= 325{,}283 \bmod 676 \\ \Rightarrow P &= 127 \end{aligned}$$

since $325{,}283 = 481 \times 676 + 127$.

Check that this is the correct P-value, either by consulting the set of P numbers above or by reenciphering it.

The simple device of partitioning plaintext into blocks consisting of two characters has greatly increased the size of the keyspace, even for a 26-letter alphabet. Using digraph encipherment with a 127-letter alphabet increases the size of the keyspace to over 258 million possibilities. I say this not to suggest that this form of enciphering is secure, but to show that some avenues of attack, such as searching a keyspace, can be made computationally prohibitive for an enemy. (In fact, because much work has been done on the frequency analysis of digraphs, such encipherment is no longer used.)

Bigger Blocks

The enciphering of the sixty-nine characters of

go placidly amid the noise and haste and
remember what peace there may be in silence

using blocks of three characters, or *trigraphs,* begins with the partition

gop lac idl yam idt hen ois ean dha ste and
rem emb erw hat pea cet her ema ybe ins ile nce

Since the total number of possible trigraphs is 26^3, or 17,576, each trigraph can be assigned a unique number from 0 to 17,575. One way of doing this is illustrated for the trigraph **gop**. Since the numerical equivalents of the single characters **g, o** and **p** are 6, 14 and 15, respectively, the P-number assigned to the trigraph **gop** is

$$(6 \times 26^2) + (14 \times 26) + 15 = 4056 + 364 + 15 = 4435$$

These are the P-numbers for all the trigraphs:

4435 7438 5497 16,236 5505 4849 9690 2717 2210 12,666 341

11,608 3017 3168 4751 10,244 1475 4853 3016
16,254 5764 5698 8844

If the same encipherment rule as already described for digraphs is used with the modulus 17,576, then the ciphertext number corresponding to the P-number 4435 is

$$C = [(159 \times 4435) + 580] \bmod 17{,}576 = 2705$$

Because

$$2705 = (4 \times 26^2) + (0 \times 26) + 1$$

and since the numbers 4, 0 and 1 are the numerical equivalents of the letters **e, a** and **b,** respectively, the encipherment gives

gop → **eab**

The complete encipherment of

go placidly amid the noise and haste and
remember what peace there may be in silence

is

eabiiotupxrsvrnxjvsakpxvaripzudbr
bdsimlrzwainsiijunyihigibxuepkpbubau

This is a completely different encipherment from the one obtained using digraphs.

The plaintext may be partitioned into blocks of length 23:

goplacidlyamidthenoisea ndhasteandrememberwhatp
eacetheremaybeinsilence

The total number of blocks of length 23 is

$$n = 26^{23} = 350{,}257{,}144{,}982{,}200{,}575{,}261{,}531{,}309{,}080{,}576$$

which is a "fairly big" number, though by comparison with the numbers used in modern cryptography it is a toddler. Assigning the three blocks a P-number in the range 0 to n using a similar procedure to those already described gives the P-numbers

88,389,774,112,496,488,148,491,229,603,504
176,823,017,496,378,363,396,475,650,125,589
53,929,205,030,218,752,977,105,934,412,508

There is a multitude of encipherment rules of the form

$$C = (mP + s) \bmod n$$

corresponding to the different choices of m and s, any of which may be chosen. One such has the enciphering key

$$K_E = \langle 162{,}547{,}535{,}981{,}244{,}558{,}392{,}877{,}030{,}695{,}465,\\ 49{,}347{,}804{,}216{,}551{,}953{,}218{,}831{,}980{,}473{,}511\rangle$$

because m = 162,547,535,981,244,558,392,877,030,695,465 has no factor other than 1, in common with n = 350,257,144,982,200, 575,261,531,309,080,576. In this case the P-numbers become the C-numbers

325,033,247,458,583,127,038,976,320,618,199
161,835,699,246,069,689,468,993,831,092,484
283,986,321,699,025,556,067,036,580,053,475

The ciphertext blocks of length 23 corresponding to these numbers are

ydigsikvpswonjdqeasgszgh maiyyygmlgalquxomyxxvua vccniaasyjkvkwlugphrimt

This time,

go placidly amid the noise and haste and
remember what peace there may be in silence

enciphers as

ydigsikvpswonjdqeasgszghmaiyyygmlg
alquxomyxxvuavccniaasyjkvkwlugphrimt

This last example helps us to appreciate how big the numbers can get when plaintext is partitioned into even moderately sized blocks. Although cryptosystems based on enciphering blocks of plaintext by the "multiply and shift" rules described are vulnerable to various forms of attack, modern cryptosystems rely in part for their security on the "blocking" of plaintext into huge blocks. When the modulus is a 200-digit number, the blocks of text can be of length 95 or more, and the number of different blocks of this length is astronomical.

I don't like having exhibited deciphering rules without explaining them. If in so doing I have made them appear mysterious but aroused your curiosity, then this is something positive. However, it may be that you are totally frustrated by not knowing how they are obtained. It is easy to determine the deciphering rule corresponding to an enciphering rule of the type described if one knows how to obtain the multiplicative inverse of a number (with respect to a modulus). Appendix C, page 360, gives a short account of a wonderfully efficient procedure known to Euclid for calculating the greatest common divisor of two natural numbers. This allows the inverse of a number to be calculated and leads to an explanation of how those deciphering rules are obtained.

I'm sure that other questions will have occurred to you, such as "How many natural numbers less than a given number are relatively prime to it?" (Am I putting words in your mouth?) As we shall see, this is a particularly important question in cryptography.

Functions: Notion and Notation

Each of the following enciphering rules

$$C = (P + 3) \bmod 26$$
$$C = 5P \bmod 26$$
$$C = 4P \bmod 26$$
$$C = (5P + 18) \bmod 26$$
$$C = (159P + 580) \bmod 676$$

transforms a set of P-numbers into a set of C-numbers. Whether or not, from the viewpoint of cryptography, these rules are chosen cleverly, their function is to convert P-numbers into C-numbers. Each is a function that describes the desired mathematical action to be performed on a typical P-number in order to generate the corresponding C-number.

Since the first letter of the word "function" is f, this letter is often used to stand for the function to be performed on a P-number to produce a C-number:

$$C = f(P)$$

read as "f of P." (The notation does not mean f multiplied by P, which would be written simply as fP, if f and P both stood for numbers.)

Although the symbol P in $f(P)$ stands for a number, the symbol f stands for the operation that is to be performed on the number P (to produce the number C). Until the function f is explicitly specified by some mathematical formula, we don't know the exact rule by which the C-numbers are calculated. We only know in abstract terms that the C-numbers are "a function of" the P-numbers. The manner in which they are computed depends on a nonspecified action f applied to the P-numbers. When the function is expressed by a formula, such as

$$f(P) = P^3$$

then the action of f on P is specified. In this case the function f takes each P and computes its "cube," P^3, to obtain the corresponding C-value. Thus if P is replaced by 2 in the formula, then

$$f(2) = 2^3 = 8$$

The effect of applying the function f to the P-number 2 is to cube it, to get the C-number 8. Also, because

$$f(5) = 5^3 = 125$$

the function f "maps" the P-number 5 onto the C-number 125. Expressed generally in words, the function f takes a number and cubes it. This particular function f could be named the "cubing function."

The five enciphering rules listed on page 107 are thus the functions

$$f(P) = (P + 3) \bmod 26$$
$$f(P) = 5P \bmod 26$$
$$f(P) = 4P \bmod 26$$
$$f(P) = (5P + 18) \bmod 26$$
$$f(P) = (159P + 580) \bmod 676$$

respectively. When the function f is the third of these, then $C = f(P)$ translates to

$$C = 4P \bmod 26$$

and for the fourth function f, $C = f(P)$ translates to

$$C = (5P + 18) \bmod 26$$

There is a world of difference between using the first and the second of these two functions as an enciphering rule in a cryptosystem. You may recall that the first one does not map the twenty-six P-numbers $\{0, 1, 2, \ldots, 24, 25\}$ onto a set consisting of twenty-six different C-numbers. Instead it maps these twenty-six P-numbers onto the set $\{0, 2, \ldots, 22, 24\}$ consisting of just thirteen different numbers. Here each C-number has two different P-numbers associated with it. Consequently, it is worthless as an enciphering function because it does not allow a C-number to be deciphered uniquely—it does not establish a one-to-one correspondence between the set of P-numbers and the set of C-numbers. In mathematical terms, it is not a *one-to-one function*.

It should be obvious that enciphering functions must be one-to-one to ensure unique decipherment. The function $f(P) = (5P + 18) \bmod 26$ is a one-to-one function and so can be used as an enciphering function. It establishes a one-to-one correspondence between the set of twenty-six P-numbers $\{0, 1, 2, \ldots, 23, 24, 25\}$ and a set of C-numbers, which also consists of twenty-six numbers;

in fact, it is the original set of numbers ordered in a different way. Under the transformation induced by this function f, each P-number is assigned a unique C-number and, crucially, each C-number is associated with one and only one P-number. With the exception of $f(P) = 4P \bmod 26$, all the other functions listed above are also one-to-one functions and so can be used as enciphering transformations.

In general, the function that undoes the action of a one-to-one function f is called its inverse function. It is written as f^{-1} (for reasons we need not go into) and is read as "f inverse." In cryptography the deciphering function is the one that undoes the (forward) action of the enciphering function. If the (one-to-one) enciphering function is denoted by f, then the corresponding deciphering function is denoted by f^{-1}. Because f^{-1} reverses, or "inverts," the action of f by mapping C-numbers (back) to P-numbers, we can write

$$P = f^{-1}(C)$$

This notation shows f^{-1} acting on C to produce P.

In most cases, the formula or rule specifying the deciphering function is different from the rule describing the enciphering function. For example, if the enciphering function is

$$f(P) = (5P + 18) \bmod 26$$

then the deciphering function is

$$f^{-1}(C) = (21C + 12) \bmod 26$$

For the sake of emphasis, let's check that f^{-1} undoes the action of f on the specific number 19. Applying the enciphering function f first gives

$$f(19) = [(5 \times 19) + 18] \bmod 26 = 113 \bmod 26 = 9$$

because $113 = 4 \times 26 + 9$. Applying the deciphering function f^{-1} to this result then gives

$$f^{-1}(9) = [(21 \times 39) + 12] \bmod 26 = 201 \bmod 26 = 19$$

because $201 = 7 \times 26 + 19$. Thus

$$f^{-1}(9) = 19$$

verifying that f^{-1} undoes the action of f on $P = 19$.

Since $f(19) = 9$, replacing the 9 in the equation $f^{-1}(9) = 19$ makes it

$$f^{-1}(f(19)) = 19$$

It is usual to replace the "outer" pair of round brackets ( ) by a pair of square brackets [ ] thus:

$$f^{-1}[f(19)] = 19$$

This can be interpreted as saying that performing the function f on 19, and following this by performing the function f^{-1} on the result, gets us back to 19.

In general, since f^{-1} undoes what f does, then following the action of f on P by that of f^{-1} on the result has the effect of reproducing P, a fact which is expressed concisely by writing

$$f^{-1}[f(P)] = P$$

In terms of enciphering and deciphering, this simply says that decrypting what has been encrypted returns the original plaintext.

A Cryptosystem

The function notation allows us to describe in simple terms the general character of a cryptosystem. Such a system essentially consists of an enciphering function, or transformation f, which does the encrypting, and a deciphering transformation f^{-1}, which does the decrypting. The equation

$$C = f(P)$$

represents f acting on a typical plaintext number P to give its corresponding ciphertext number C, while the equation

$$P = f^{-1}(C)$$

represents f^{-1} acting on a typical ciphertext number C to restore the original plaintext number P.

Usually a cryptosystem is thought of not as one enciphering function along with its associated deciphering function, but as a family of enciphering functions and corresponding deciphering functions. For example, for a fixed modulus n, each member f of the family of "multiply and shift" enciphering transformations has the form

$$C = f(P) = (mP + s) \bmod n$$

for a particular pair of values of the parameters m and s. The number of different enciphering functions is determined by the number of distinct pairs $\langle m, s\rangle$. If this "cryptosystem" were to be used as a general system by a number of people, then each user would have his or her own special enciphering key, $K_E = \langle m, s\rangle$. This key has a unique deciphering key, K_D, associated with it, which determines the corresponding deciphering transformation f^{-1}.

According to Kerckhoffs's principle, the users of such a system should assume that the general method by which they encipher messages is public knowledge. Their confidence in whether or not the system is any good should depend only on believing that keeping the keys secret ensures secure communication. Apparently, in this case, it would be a mistaken belief even for a very large modulus.

Because the letters P and C are the initials of the words "plaintext" and "ciphertext," respectively, it is natural for the enciphering and deciphering functions to be described by the equations

$$C = f(P)\ ;\ P = f^{-1}(C)$$

However, when the letters *P* and *C* are replaced by the more commonly used *x* and *y*, the equations read

$$y = f(x)\ ;\ x = f^{-1}(y)$$

This familiar notation will be used later when we discuss the remote coin-tossing protocol used by two of cryptography's best-known personages, Alice and Bob.

7 Sums with a Difference

In this chapter you get a chance to learn more about the powerful arithmetic used in cryptography which I myself began to learn at my father's evening class—and which I truly began to admire when I had to implement some of its features while I was taking my first steps in programming. This arithmetic can give answers to problems in just two or three steps where ordinary arithmetic would require massive amounts of calculation. Along the way we'll learn about the rule of nines, how books are assigned ISBNs, googols, a mistaken belief of the ancient Chinese, pseudoprimes, Carmichael numbers and the Little Theorem of Fermat. We'll discover the almost incredible fact that a number can be declared to be composite without a single factor having been found. I'll tell you a little about the problem of *primality* (finding primes), and its sister problem, *factoring*.

More on Days of the Week

Let's return to the problem in which we assume that today is Sunday and are interested in knowing what day of the week it will be one hundred days from now. Suppose we assign the number 0 to this Sunday. Then to tomorrow, Monday, we assign the number 1, to Tuesday 2, and so on to next Saturday, which is assigned the number 6. Thus we have the following correspondences between the numbers 0, 1, 2, 3, 4, 5, 6 and the days of this week, which starts today, Sunday:

Sunday	$\leftrightarrow$	0
Monday	$\leftrightarrow$	1
Tuesday	$\leftrightarrow$	2
Wednesday	$\leftrightarrow$	3
Thursday	$\leftrightarrow$	4
Friday	$\leftrightarrow$	5
Saturday	$\leftrightarrow$	6

It is then natural that all future Sundays be assigned the numbers 7, 14, 21, . . . in succession. We can even assign numbers to all past Sundays by labeling last Sunday with the negative number -7, the Sunday before that with the number -14, and so on for all the Sundays receding into the far-off past. Thus the infinite set of integers (whole numbers)

$$\{\ldots, -21, -14, -7, 0, 7, 14, 21, \ldots\}$$

contains the numerical labels of all Sundays, past, present and future. We could call this set "the set of Sundays." By the same token, the following infinite sets of integers contain the numerical labels of all past and future

Mondays:	$\{\ldots, -20, -13, -6, 1, 8, 15, 22, \ldots\}$
Tuesdays:	$\{\ldots, -19, -12, -5, 2, 9, 16, 23, \ldots\}$
Wednesdays:	$\{\ldots, -18, -11, -4, 3, 10, 17, 24, \ldots\}$
Thursdays:	$\{\ldots, -17, -10, -3, 4, 11, 18, 25, \ldots\}$
Fridays:	$\{\ldots, -16, -9, -2, 5, 12, 19, 26, \ldots\}$
Saturdays:	$\{\ldots, -15, -8, -1, 6, 13, 20, 27, \ldots\}$

These sets have no element in common (how could the set of Mondays, say, have an element in common with the set of Fridays?), but between them they give the full set of integers

$$\{\ldots, -4, -3, -2, -1, 0, 1, 2, 3, 4, \ldots\}$$

This is hardly a surprise since the number 0 is associated with today, the positive integers with all future days and the negative integers with all the days that have passed. So every integer lies in one of these sets, and in one set only.

All the numbers in the Tuesdays-set

$$\{\ldots, -19, -12, -5, 2, 9, 16, 23, \ldots\}$$

leave the remainder 2 when divided by 7, even the negative ones. For example, since -19 can be written as $-3 \times 7 + 2$, the remainder when -19 is divided by 7 is 2. Although -19 is also equal to $-2 \times 7 + (-5)$, the number -5 is not regarded as the remainder of -19 upon division by 7. As we've seen, the convention with respect to a remainder is that it must be either 0 or a positive integer less than the modulus. For division by 7 this means that it is one (and only one) of 0, 1, 2, 3, 4, 5, 6. The fact that every integer in the Tuesdays-set leaves a remainder of 2 when divided by 7 may strike you as nothing more than obvious. The technical term for this set is "the residue class 2 modulo 7." The word "residue" is just another word for "remainder," and "class" is just another word for "set."

Similarly, every integer in the Wednesdays-set leaves a remainder of 3 when divided by 7; this is the residue class 3 modulo 7. Similar results and names apply to the other sets. So if you need to know to which residue class a given integer belongs, all you have to do is find the remainder obtained when the number is divided by the modulus 7, or, as we say more briefly, its "remainder modulo 7." The calculation

$$100 = 14 \times 7 + 2 \Rightarrow 100 \bmod 7 = 2$$

tells us that the number 100 lies in the residue class 2 modulo 7.

Two integers, call them a and b, belong to the same residue class modulo 7 if and only if

$$a \bmod 7 = b \bmod 7$$

This is one way of saying that they both leave the same remainder upon division by 7. Thus, to find whether 675 and 943 lie in the same residue class modulo 7, calculate 675 mod 7 and 943 mod 7 to see whether these two numbers match. Of course, in the context of weekdays, lying in the same class means that both integers represent the same day of the week. Since 675 mod 7 = 3 and 943 mod 7 = 5, we see that 675 days from now it will be a Wednesday, while 943 days from now it will be a Friday. Thus 675 and 943 lie in different residue classes modulo 7.

Whereas finding their respective remainders allows us to determine the days of the week associated with two numbers, there is a much faster way of deciding whether two integers belong to the same residue class. It is clear that the integers in any particular class differ from one another by a multiple of 7 since, for example, all Tuesdays are "full weeks" apart. This means that for any two integers, a and b, to be in the same residue class, their difference, $a - b$, should be exactly divisible by 7.

What Class Am I In?

We write $m|n$ (and say "m divides n") when the integer m divides exactly into the integer n and write $m \nmid n$ (and say "m does not divide n") when it doesn't. So, for example, $7|21$ but $7 \nmid 22$. Similarly, $11|132$ while $15 \nmid 26$.

So, if two integers, a and b, are such that $7|(a - b)$ (meaning that their difference, $a - b$, is exactly divisible by 7), then they lie in the same residue class modulo 7. It is customary to write $7|a - b$ rather than $7|(a - b)$. If $7 \nmid a - b$, then their difference is not divisible by 7, and a and b lie in different residue classes. For example, to show that the numbers 855 and 127 lie in the same residue class, we calculate their difference 855 − 127 to be 728 and then check whether $7|728$. It does because $728 = 104 \times 7$.

In terms of days of the week, the numbers 127 and 855 represent the same weekday. Note that though the above calculation tells us that these two weekdays lie a full 104 weeks apart, it does not indicate which day of the week they both represent. This extra information is obtained by working out that 127 mod 7 = 1 or 855 mod 7 = 1. This shows that each of them represents a Monday.

On the other hand, the numbers 943 and 675 lie in different sets because their difference, 943 − 675 = 268, is not divisible by 7, as the calculation 268 = 38 × 7 + 2 shows. This calculation gives no indication of the different days represented by 943 and 675, but it does reveal that they represent weekdays that are two days apart. If the number 675 represents a Wednesday, which day of the week does the number 943 represent? (See Appendix B, page 302, to check your answer.)

Karl Friedrich Gauss introduced the notation

$$a \equiv b \pmod{m}$$

to indicate that the integers a and b lie in the same residue class modulo m. (Read $a \equiv b \pmod{m}$ as "a is congruent to b, modulo m.") Gauss deliberately used the symbol ≡ because of its resemblance to the equality symbol =. He developed a new arithmetic similar to ordinary arithmetic, which is ideally suited for calculating with remainders.

The letter m, standing for the modulus, is always a natural number. So far we have been working with the modulus $m = 7$. For example, $38 \equiv 17 \pmod{7}$ because 38 and 17 both lie in the residue class 3 modulo 7. Thus we say that 38 is congruent to 17 modulo 7. We also say that 38 and 17 are congruent modulo 7.

The notation $a \not\equiv b \pmod{m}$ can be read as "a is not congruent to b, modulo m." In other words, a and b do not lie in the same residue class modulo m. In this case a and b are said to be incongruent modulo m, so $a \not\equiv b \pmod{m}$ is also read as "a and b are incongruent modulo m." For example, $24 \not\equiv 16 \pmod{5}$ because 24 lies in the residue class 4 modulo 5, while 16 lies in the residue class 1 modulo 5. Thus 24 and 16 are incongruent modulo 5.

You may have noticed that the symbol "mod" is now being used in a different role, and appears between brackets followed by the modulus m. In our earlier use it appears without brackets, followed by the modulus m. However, there is a connection between the two uses.

Whereas the expression 100 mod 7 is the remainder when 100 is divided by 7, the expression $100 \equiv 86 \pmod 7$ says that 100 and 86 lie in the same residue class modulo 7. But these numbers lie in the same residue class modulo 7 if and only if they leave the same remainder when divided by 7—that is, if and only if

$$100 \bmod 7 = 86 \bmod 7$$

They are, since both are 2. In general,

$$a \equiv b \pmod m \text{ if and only if } a \bmod m = b \bmod m$$

This relation displays the two different uses of the mod symbol. It also tells us that if we want to check whether $a \equiv b \pmod m$, then all we need do is check whether $a \bmod m = b \bmod m$.

But you may recall from our discussion of residue classes modulo 7 that there is a far easier way. To check whether $a \equiv b \pmod m$, all you need do is check whether their difference, $a - b$, is divisible by m, since then and only then do a and b lie in the same residue class modulo m. Thus

$$a \equiv b \pmod m \text{ if and only if } m|a - b$$

This criterion is the simplest way of checking whether $a \equiv b \pmod m$.

We can confirm that $38 \equiv 17 \pmod 7$ by showing that the difference $38 - 17 = 21$ is divisible by 7. But $7|21$ because $21 = 3 \times 7$. Normally we are not as long-winded as this. To check whether $38 \equiv 17 \pmod 7$ we simply write $7|38 - 17 = 21$. The expression $38 - 17 = 21$ to the right of the divisibility symbol $|$ is the difference between 38 and 17, being calculated as 21. Then the expression is interpreted as reading $7|21$. On the other hand,

$24 \not\equiv 16 \pmod 7$ because $24 - 16 = 8$ is not divisible by 7. As above, we might simply write $24 \not\equiv 16 \pmod 7$ because $7 \nmid 24 - 16 = 8$.

Now for a short tour of some of the features of this arithmetic. (A note on terminology: $32 \equiv 14 \pmod 9$ is often termed a "congruence with respect to the modulus 9," or a "congruence modulo 9," or simply a "congruence" when the modulus is understood.)

Arithmetic Properties of Congruences

Multiplication by a Constant

A congruence can be multiplied on both sides by the same whole number to get another valid congruence.

Multiplying both sides of the congruence $32 \equiv 14 \pmod 9$ by 8 gives

$$32 \times 8 \equiv 14 \times 8 \pmod 9$$

Performing ordinary multiplication on both sides of the congruence gives

$$256 \equiv 112 \pmod 9$$

Check: Since $9|256 - 112 = 144$, this congruence is true.

Exponentiation

Both sides of a congruence can be raised to the same positive number (often called the power, or exponent). This is called *exponentiating* the congruence.

For example, raising each side of the congruence

$$32 \equiv 14 \pmod 9$$

to the power 3 gives

$$32^3 \equiv 14^3 \pmod 9$$

This says that

$$32{,}768 \equiv 2{,}744 \pmod 9$$

Check: Since $32{,}768 - 2{,}744 = 30{,}024 = 3{,}336 \times 9$, this is a valid congruence.

Addition, Subtraction and Multiplication

Two congruences *with respect to the same modulus* can be "added," "subtracted" and "multiplied." This is illustrated by the two congruences

$$32 \equiv 14 \pmod 9 \; ; \; 25 \equiv 7 \pmod 9$$

These congruences are "added" as follows:

$$\begin{array}{rcr} 32 & \equiv & 14 \pmod 9 \\ \underline{25} & \underline{\equiv} & \underline{7 \pmod 9} \\ 57 & \equiv & 21 \pmod 9 \end{array}$$

Check: Here, ordinary addition is carried out on the left- and right-hand sides to give the single congruence

$$57 \equiv 21 \pmod 9$$

Is this a true statement? Yes, because $9 | 57 - 21 = 36$. Great!

The same two congruences are "subtracted" as follows:

$$\begin{array}{rcr} 32 & \equiv & 14 \pmod 9 \\ \underline{25} & \underline{\equiv} & \underline{7 \pmod 9} \\ 7 & \equiv & 7 \pmod 9 \end{array}$$

Check: Here, ordinary subtraction is carried out on the left- and right-hand sides to give the single congruence:

$$7 \equiv 7 \pmod 9$$

This is also true because $9 \mid 7 - 7 = 0$.

The same two congruences are "multiplied" as follows:

$$\begin{array}{rcr} 32 & \equiv & 14 \pmod 9 \\ 25 & \equiv & 7 \pmod 9 \\ \hline 800 & \equiv & 98 \pmod 9 \end{array}$$

Here, ordinary multiplication is carried out on the left- and right-hand sides to give the single congruence

$$800 \equiv 98 \pmod 9$$

Check: Since $800 - 98 = 702 = 78 \times 9$, this congruence is also valid.

We don't talk (at this stage at any rate) about "dividing" congruences since we are dealing only with whole numbers and not fractions.

The properties illustrated for the modulus 9 are also valid for any other natural number modulus. You might like to make up your own examples using a different modulus.

The Transitive Property

In ordinary arithmetic this property says that if $a = b$ and $b = c$, then $a = c$, something which may strike you as obvious. This property is also valid in our new arithmetic. For example,

$$\text{if } 147 \equiv 84 \pmod 9 \text{ and } 84 \equiv 3 \pmod 9 \text{ then } 147 \equiv 3 \pmod 9$$

Let us check that all these congruences are true. Since $147 - 84 =$

63 and 63 is divisible by 9 the first congruence is true. The second is also true because $84 - 3 = 81 = 9 \times 9$. The last is true because $147 - 3 = 144 = 16 \times 9$.

The first congruence says that 147 and 84 are in the same residue class modulo 9, while the second says that 84 and 3 are in the same residue class modulo 9. Hence it is obvious that 147 and 3 are also in the same residue class modulo 9, or equivalently that $147 \equiv 3 \pmod 9$.

A Million Days from Now

To determine which day of the week it will be a million days from now, begin with

$$10 \equiv 3 \pmod 7$$

Now square both sides of this congruence to get

$$100 \equiv 9 \pmod 7$$

Since $9 \equiv 2 \pmod 7$, the transitive property says that

$$100 \equiv 2 \pmod 7$$

Since 100 cubed is 1,000,000, cube both sides of this congruence to get

$$1{,}000{,}000 \equiv 8 \pmod 7$$

Since $8 \equiv 1 \pmod 7$, the transitive property gives

$$1{,}000{,}000 \equiv 1 \pmod 7$$

Thus the number 1,000,000 is in the residue class 1 modulo 7. In the context of days of the week, if today is Sunday then in one million days it will be a Monday. We have arrived at this result almost effortlessly, by using some properties of congruences along with a few simple multiplications and divisions.

However, you might say, "Wouldn't it be easier (if less elegant) to find 10^6 mod 7 simply by using short division to get

$$7\overline{)1{,}000{,}000}$$
$$142{,}857 + 1$$

This calculation tells me that a million days consist of 142,857 weeks with a day to spare. It may appear to be just as easy to tackle the problem this way because dividing by 7 isn't that hard, even though it does involve six divisions and six "carries." But I'll say to you, "OK, smarty, which day of the week will it be in 10^{100} days' time?"

Googol

The number 10^{100} has one hundred and one digits, beginning with 1 and followed by a hundred 0's. It is known as a googol, a name invented by a nine-year-old American boy, Milton Sirotta, nephew of mathematician Edward Kasner. Numbers of this magnitude are commonplace in cryptography.

Knowing the congruence $1{,}000{,}000 \equiv 1 \pmod 7$ allows us to show that the googol, 10^{100}, leaves a remainder of 4 when divided by 7. This is done without knowing how often 7 divides into this gigantic number. Here's how:

Since $1{,}000{,}000 = 10^6$, we know from the "million days from now" calculation that

$$10^6 \equiv 1 \pmod 7$$

The 1 on the right-hand side of this congruence is very convenient. Why? Because no matter to what power 1 is raised it still remains a 1. Watch how this property is exploited. We raise both sides of the congruence to the power of 16 because $(10^6)^{16} = 10^{96}$ (note that $(10^6)^{17} = 10^{102}$ makes the power exceed 100). We get

$$10^{96} \equiv 1 \pmod 7$$

We're nearly there. To finish, we need to multiply 10^{96} by 10^4 to get 10^{100}. Since $10^2 \equiv 2 \pmod 7 \Rightarrow 10^4 \equiv 4 \pmod 7$, the multiplication

$$\begin{array}{lcl} 10^{96} & \equiv & 1 \pmod 7 \\ \underline{10^4} & \underline{\equiv} & \underline{4 \pmod 7} \\ 10^{100} & \equiv & 4 \pmod 7 \end{array}$$

shows that

$$10^{100} \equiv 4 \pmod 7$$

So if today is a Sunday, then in 10^{100} days it will be Thursday. You shouldn't go through life without knowing this! On a more serious note, raising one number, such as 10 here, to another number, such as 100 here, is of central importance in cryptography. However, the numbers used are enormous by comparison with the base number 10 and the exponent 100. Typically the base is a 200-digit number and the exponent can have as many as 100 digits. The problem of raising the base number to the power of the exponent is called the *exponentiation problem.* As I shall explain later it is the key step in the wonderful RSA cryptosystem but is, unfortunately and as you might suspect, a time-consuming step which renders the RSA system slower than one wants it to be. My second project investigated a different type of system that tries to avoid this step in order to achieve faster encipherment and decipherment. But all this is a story for later. For the moment I want to show you one or two more things to make you more familiar with working with congruence arithmetic. While they have no direct relevance to cryptography, they are not without interest or application.

A Special Property of the Number 9

We'll put the new arithmetic to work to show you a property of the number 9. So for the next few pages we'll be working with the modulus 9.

Beginning with the fact that

$$10 \equiv 1 \pmod 9$$

and squaring both sides of the congruence—that is, raising each side to the power 2—gives

$$10^2 \equiv 1 \pmod 9$$

Is this true? To show that $10^2 \equiv 1 \pmod 9$ we must verify that $9 | 10^2 - 1$. Since $10^2 - 1 = 100 - 1 = 99$, this is clearly the case.

Cubing both sides of the original congruence gives

$$10^3 \equiv 1 \pmod 9$$

This is also true since $10^3 - 1 = 1000 - 1 = 999$ is divisible by 9. Similarly,

$$10^4 \equiv 1 \pmod 9$$

Check: $10^4 - 1 = 10{,}000 - 1 = 9999$ is divisible by 9. By now you will have guessed that it doesn't matter to what power 10 is raised, the result is always congruent to 1 modulo 9. That is,

$$10^n \equiv 1 \pmod 9$$

for each natural number n. It is also true for $n = 0$ because $10^0 = 1$. Put another way, this says that $1 \equiv 1 \pmod 9$.

If, for the following four congruences modulo 9,

$$\begin{aligned} 1000 &\equiv 1 \pmod 9 \\ 100 &\equiv 1 \pmod 9 \\ 10 &\equiv 1 \pmod 9 \\ 1 &\equiv 1 \pmod 9 \end{aligned}$$

we multiply the first by 3, the second by 5, the third by 4 and the fourth by 8, then we get

$$\begin{aligned} 3000 &\equiv 3 \pmod 9 \\ 500 &\equiv 5 \pmod 9 \\ 40 &\equiv 4 \pmod 9 \\ 8 &\equiv 8 \pmod 9 \end{aligned}$$

Since these congruences are all modulo 9 they can be added (on each side) to give

$$3548 \equiv 3 + 5 + 4 + 8 \pmod 9$$

We have deliberately refrained from actually adding $3 + 5 + 4 + 8$ to get 20. Now that we have done so, we can verify that the congruence is true because $3548 - 20 = 3528 = 392 \times 9$.

So where is all this going? We have discovered that the number 3548 is congruent to "the sum of its digits" modulo 9. This is true of every natural number n. In general, if s is the sum of the digits representing n, then

$$n \equiv s \pmod 9$$

This means that n and s always lie in the same residue class modulo 9, or equivalently that $9 | n - s$.

If $n = 83{,}725{,}163$, then $s = 8 + 3 + 7 + 2 + 5 + 1 + 6 + 3 = 35$ and

$$n - s = 83{,}725{,}163 - 35 = 83{,}725{,}128 = 9{,}302{,}792 \times 9$$

shows that $9 | n - s$, or equivalently that $n \equiv s \pmod 9$.

Now for a small piece of cleverness which will tell us in which residue class modulo 9 both n and s lie. Applying the same result to $s = 35$ gives

$$35 \equiv 3 + 5 = 8 \pmod 9$$

Thus $s \bmod 9 = 8$, and so $n \bmod 9 = 8$ also. Hence 83,725,163 and 35 both lie in the residue class 8 modulo 9. It is easy to verify this result for 35 because $35 = 3 \times 9 + 8$.

The Rule of Nines

You might remember this rule, which says that an integer is divisible by 9 if and only if the sum of its digits is divisible by 9.

Here is the explanation. If the sum s of the digits of a number is divisible by 9, then $s \equiv 0 \pmod 9$. Since $n \equiv s \pmod 9$ in general, in this particular case $n \equiv 0 \pmod 9$ also. But $n \equiv 0 \pmod 9$ means that the number n itself is divisible by 9.

Knowing this rule might get you looking at license plate numbers to see whether or not they are divisible by 9. When adding up the digits you can save work by continually reducing modulo 9, or "casting out nines" as it is otherwise known. Suppose, for example, you see the number 83,725,163 (which is a very long license plate!) and you start to add its digits from the left. When you add 8 to 3 you get 11, but you must take it to be 2 since $11 \equiv 2 \pmod 9$. You have reduced 11 to 2, modulo 9, by casting out a nine. Now adding this 2 to the next digit, 7, gives 9, but reducing modulo 9 it becomes 0. You have just cast out another nine. Now you can start again at the digit 2, and add it to the next digit, 5, to get 7, and add this to 1 to get 8. Then add this 8 to 6 to get 5. (To find out why, see Appendix B, page 302.) Finally you add this 5 to 3 to get 8. You conclude that $83{,}725{,}163 \equiv 8 \pmod 9$. Since this remainder is not 0 you now know that the integer 83,725,163 is not divisible by 9.

You can take all sorts of liberties when you are determining $s \bmod 9$, as shown in the following variation of the calculation just described:

$$\begin{aligned} s &= 8 + 3 + 7 + 2 + 5 + 1 + 6 + 3 \\ &= (8 + 3) + (7 + 2) + (5 + 1) + (6 + 3) \\ &\equiv 2 + 0 + 6 + 0 \pmod 9 \\ \Rightarrow s &\equiv 8 \pmod 9 \\ \Rightarrow s \bmod 9 &= 8 \end{aligned}$$

There are many rules known for deciding whether or not an integer is divisible by small integers such as 4, 7 and 11, to name but a few, but none is as simple or as elegant as the one relating to divisibility by 9. The one that tells when an integer is divisible by 11 is fairly easy to describe, but I would rather show a more "worldly" application of the modulus 11.

Book Numbering

Every modern book carries a ten-digit number on its back cover, normally near the bottom right-hand corner. This is true of the Irish/English hardback edition of this book, which carries the ten-digit number 1 86197 222 9. This number is the book's ISBN (International Standard Book Number). Its first nine digits are "information digits" because they code information about the book; in this case the first digit, 1, identifies the book as being published in an English-speaking country, and the next eight digits identify the publishing house (among other things). The tenth and final digit is known as a *check digit,* and though it is redundant in terms of information, it is very useful in checking whether or not a purported ISBN is a valid number.

Apparently the two types of error most commonly committed by humans when transcribing long strings of digits are a single transcription error in any position and an error due to the transposition of any two digits.

If either type of error occurs during the transcription, then the tenth digit calculated by a recipient of this supposed ISBN will not match the tenth digit of the submitted number. Thus a bookseller, for example, need not waste time searching for a book using an invalid ISBN. Let me describe to you how this check digit is calculated, without explaining why computing it in this way provides a "flag" for the presence of one of the errors just mentioned.

The first information digit is multiplied by 1 (not very difficult), the second by 2, the third by 3, and so on, to the ninth digit by 9. The results of all these multiplications are added up to get what is known as the *weighted check sum,* say, S. Then the tenth check digit is taken to be S mod 11. That's all there is to it (which is not quite true because you can take some shortcuts in calculating S).

To calculate the check digit in the ISBN for the book whose first nine information digits are 1 86197 222, compute

$$\begin{aligned} S &= (1 \times \mathbf{1}) + (2 \times \mathbf{8}) + (3 \times \mathbf{6}) + (4 \times \mathbf{1}) + (5 \times \mathbf{9}) + (6 \times \mathbf{7}) \\ &\qquad + (7 \times \mathbf{2}) + (8 \times \mathbf{2}) + (9 \times \mathbf{2}) \\ &= 1 + 16 + 18 + 4 + 45 + 42 + 14 + 16 + 18 \\ &= 174 \end{aligned}$$

Then S mod 11 = 9, since $174 = 15 \times 11 + 9$. Appending the check digit 9 to the nine information digits gives the ISBN as 1 86197 222 9.

The use of the modulus 11 in the ISBN scheme is not arbitrary; it exploits the fact that 11 is a prime number. The scheme does not work with 10 as a modulus. Since it is possible to get a remainder of 10 when working with the modulus 11, the letter X is appended instead of the two-digit 10 when this happens, so as to maintain a ten-digit ISBN. *Chamber's Twentieth Century Dictionary,* for example, has the ISBN 0 550 10206 X. The calculation

$$\begin{aligned} S &= (1 \times \mathbf{0}) + (2 \times \mathbf{5}) + (3 \times \mathbf{5}) + (4 \times \mathbf{0}) + (5 \times \mathbf{1}) \\ &\qquad + (6 \times \mathbf{0}) + (7 \times \mathbf{2}) + (8 \times \mathbf{0}) + (9 \times \mathbf{6}) \\ &\equiv 0 + (-1) + 4 + 0 + 5 + 0 + 3 + 0 + \\ &\qquad (-1) \pmod{11} \\ \Rightarrow S &\equiv 10 \pmod{11} \\ \Rightarrow S \bmod 11 &= 10 \end{aligned}$$

verifies that this is so. Note the clever shortcuts based on casting out 11's and using negative numbers (e.g., -1 instead of 10 and 54) to keep computation to a minimum. Any trick or shortcut that can be applied in general to speed up a calculation is invaluable in cryptography because fast encipherment and decipherment are often crucial, particularly in military operations.

Fermat's Little Theorem

The name Fermat has become known to many who might otherwise never have heard of him through Simon Singh's book *Fermat's Last Theorem.* Its main story is how the mathematician Andrew Wiles proved this famous 350-year-old result in 1994. The adjective "last" was used because this conjecture was the last of Fermat's many assertions that mathematicians had still not verified. His many other statements about natural numbers—and there were many—were in the course of time proved to be true, with the exception of just a few, which were shown by counter-examples to be false. However, this old chestnut defeated the best attempts of the greatest mathematicians over the centuries until at last the Englishman (who had been captivated from childhood by the problem) devoted almost eight years, locked away in his attic, to proving it true.

But Fermat is also known for a theorem which is called *Fermat's Little Theorem,* or FLT for short. Whereas the Last Theorem could be described as a negative result in that it states that something can never happen, the FLT is a real extrovert in that it makes very positive assertions. It is also a great theorem and, in stark contrast to the Last Theorem, is much easier to prove. Dad showed us a gem of a proof from 1801, by the Scottish mathematician James Ivory.

Fermat is known to have written at some time or other:

> And perhaps posterity will thank me for having shown it that the ancients did not know everything.

Well, the mathematicians of antiquity did not know this result, and posterity is grateful to him for it. Like a lot of mathematical work which seems to have no "practical application" at the time of its creation, the FLT of 1640, along with Euler's generalization of it in 1736, became of central practical importance much later when, in 1977, they were pivotal results in establishing the soundness of the RSA public key cryptosystem.

The historical chain

Fermat →	Euler →	Rivest, Shamir and Adleman
1640	1736	1977

links different times and different people, and shows how mathematicians build on previous results.

By now you're probably thinking, Yes, yes, that's all well and good. But what does the FLT *say*?

Fermat's Little Theorem: If p is *any* prime number, then

$$a^{p-1} \equiv 1 \pmod{p}$$

for every natural number a relatively prime to p.

Before saying another word, let us choose the prime p to be the number 5. Since $p - 1 = 4$, the FLT says, in this specific case, that

$$a^4 \equiv 1 \pmod{5}$$

for every natural number a relatively prime to 5.

You'll recall that two natural numbers are relatively prime to each other if (and only if) the only factor they have in common is 1. Which natural numbers a are relatively prime to 5? First, each of the numbers 1, 2, 3 and 4. However, the number 5 shares the factor 5 with itself, so it is not relatively prime to itself. OK, so 5 cannot be an a-number. The numbers 6, 7, 8 and 9 are each relatively prime to 5, but 10 is not because 10 and 5 share the factor 5. The numbers 11, 12, 13 and 14 are also relatively prime to 5, but 15 is not because 15 and 5 share the factor 5.

I hear you yelling, "Stop! The only numbers not relatively prime to 5 are the multiples of 5!"

"Yes, the numbers 5, 10, 15, 20, . . . are the only ones sharing a factor other than 1 with 5."

So we may say that

$$a^4 \equiv 1 \pmod 5$$

for every natural number a not a multiple of 5. So the theorem asserts that

$$\begin{aligned}
1^4 &\equiv 1 \pmod 5 \\
2^4 &\equiv 1 \pmod 5 \\
3^4 &\equiv 1 \pmod 5 \\
4^4 &\equiv 1 \pmod 5 \\
6^4 &\equiv 1 \pmod 5 \\
7^4 &\equiv 1 \pmod 5 \\
8^4 &\equiv 1 \pmod 5 \\
9^4 &\equiv 1 \pmod 5 \\
11^4 &\equiv 1 \pmod 5 \\
&\vdots
\end{aligned}$$

The dots at the end of this array signify that the columns go on forever in this manner. The only values of a which do not appear are the multiples of 5. For example, there isn't an entry asserting that $20^4 \equiv 1 \pmod 5$, which is just as well because this is not a true statement. Thus the FLT makes an awful lot of assertions for just one prime. Let's verify some of them.

The first assertion is trivially true as it says that $1 \equiv 1 \pmod 5$: a number is always congruent to itself no matter what the modulus. Is it true that

$$2^4 \equiv 1 \pmod 5?$$

Yes. The assertion that $2^4 \equiv 1 \pmod{5}$ means that 5 divides the difference $2^4 - 1$. This is true because $2^4 - 1 = 16 - 1 = 15 = 3 \times 5$. Is this next statement true?

$$3^4 \equiv 1 \pmod{5}$$

It is because $3^4 \equiv 1 \pmod{5}$ means that $5 | 3^4 - 1 = 80$.

Since, for example, $3^4 \equiv 1 \pmod{5}$ is equivalent to saying that $5 | 3^4 - 1$, let us rewrite the above assertions in this form. Here is the previous array in different clothes:

$$5 | 1^4 - 1$$
$$5 | 2^4 - 1$$
$$5 | 3^4 - 1$$
$$5 | 4^4 - 1$$
$$5 | 6^4 - 1$$
$$5 | 7^4 - 1$$
$$5 | 8^4 - 1$$
$$5 | 9^4 - 1$$
$$5 | 11^4 - 1$$
$$\vdots$$

To verify these statements we need to show that each of these "differences" is a multiple of 5. In this case, all we need to do is show that they end in either a 0 or a 5. Here are the results:

$$\begin{aligned}
1^4 - 1 &= & 0 &= & 0 \times 5 \\
2^4 - 1 &= & 15 &= & 3 \times 5 \\
3^4 - 1 &= & 80 &= & 16 \times 5 \\
4^4 - 1 &= & 255 &= & 51 \times 5 \\
6^4 - 1 &= & 1295 &= & 259 \times 5 \\
7^4 - 1 &= & 2400 &= & 480 \times 5 \\
8^4 - 1 &= & 4095 &= & 819 \times 5
\end{aligned}$$

$$9^4 - 1 = 6560 = 1312 \times 5$$
$$11^4 - 1 = 14640 = 2928 \times 5$$
$$\vdots$$

Why don't you test out the FLT for the prime $p = 7$?

Since the natural numbers a which are relatively prime to a prime number p are precisely those natural numbers which are not multiples of p, the general FLT may be stated in the form:

If p is any prime number then

$$a^{p-1} \equiv 1 \pmod{p}$$

for every natural number a that is not a multiple of p.

It may also be written without the use of the congruence notation in the form

$$p | a^{p-1} - 1$$

for every natural number a that is not a multiple of p.

The FLT is a deep result: It makes an infinite number of statements with regard to each prime. Since there is an infinite number of primes, it makes "a lot" of statements, all of which are true. Let's look at some of the consequences of this wonderful theorem.

An Ancient Chinese Belief (fifth century BC)

The ancient Chinese believed that

$$\text{if } n \text{ is a prime then } n | 2^{n-1} - 1$$

and conversely that

$$\text{if } n \mid 2^{n-1} - 1 \text{ then } n \text{ is a prime}$$

(A small point: the prime must be odd, which is the case for all primes except 2.)

What is being said? The first statement says that if n is a prime then it *always* divides evenly into $2^{n-1} - 1$. For example, because 7 is a prime it must be that 7 divides evenly into $2^{7-1} - 1 = 2^6 - 1 = 63$. It does.

The second statement says that if n divides evenly into $2^{n-1} - 1$ then n is a prime.

So, for example, if n is the composite number 8, could it be that 8 divides evenly into $2^{8-1} - 1 = 2^7 - 1 = 127$? Not according to this statement. If 8 did divide evenly into 127 (which it doesn't), this would mean that 8 is a prime number. So this second statement can be viewed as saying that if n is a composite number then n *never* divides evenly into $2^{n-1} - 1$.

These two statements are often written jointly as

$$n \text{ is prime if and only if } n \mid 2^{n-1} - 1$$

Were the ancient Chinese right?

The first statement that if n is a prime then $n \mid 2^{n-1} - 1$ is true. Why? Because it's equivalent to the assertion that if n is prime, then

$$2^{n-1} \equiv 1 \pmod{n}$$

Does this look familiar? It should—it's just the FLT with $p = n$ and $a = 2$. (Since n is odd, 2 is relatively prime to it.)

So they were right in this regard. You could even go so far as to say that they knew the FLT for the case $a = 2$. Fermat's result, however, is much stronger because it says that when n is a prime, a can be *any* number other than a multiple of n.

Unfortunately, the second statement

$$\text{if } n \mid 2^{n-1} - 1 \text{ then } n \text{ must be a prime}$$

is false, in general. Amazingly, it is true for all the 271 composite numbers up to and including 340. So the ancient Chinese were right in asserting that if n is one of these composite numbers then it *never* divides evenly into $2^{n-1} - 1$. None of the composite numbers n less than or equal to 340 makes $2^{n-1} \equiv 1 \pmod{n}$. Here are some examples for composite numbers n in this range:

$$
\begin{aligned}
2^{3} &\equiv 0 \pmod{4} \\
2^{17} &\equiv 14 \pmod{18} \\
2^{44} &\equiv 31 \pmod{45} \\
2^{99} &\equiv 88 \pmod{100} \\
2^{160} &\equiv 156 \pmod{161} \\
2^{215} &\equiv 176 \pmod{216} \\
2^{339} &\equiv 8 \pmod{340}
\end{aligned}
$$

Notice that none of the numbers to the right of the $\equiv$ symbols is a 1. The ancient Chinese believed that the number 1 never appears on this side if n is composite. However, in 1820 someone discovered (who or how, I don't know) that for the composite number $341 = 11 \times 31$, it happens that

$$2^{340} \equiv 1 \pmod{341}$$

or equivalently that $341 | 2^{340} - 1$. This was no mean feat considering that the number $2^{340} - 1$ is an integer consisting of 2860 digits. Of course, the discoverer *did not* write out this monster number in full and painstakingly divide it by 341 to find that 341 divided into it evenly. We'll show how this can be done quite easily in a moment.

This unwelcome fact destroys another beautiful hypothesis. Had the Chinese been right, the fact that $341|2^{340} - 1$ would prove that 341 is a prime. Unfortunately, it is not. The number $n = 341$ is the *first* counter-example to the conjecture that

$$n | 2^{n-1} - 1 \Rightarrow n \text{ an odd prime}$$

Considering the enormity of the number $2^{340} - 1$, and the fact that the conjecture holds for 271 composite numbers less than 341, it is not hard to see why the ancient Chinese were led to believe what they did. This example serves as another lesson that one cannot argue from the "particular to the general." Although all the swans I have ever seen were white, it is not true that all swans are white. To establish a general result you have to be able to supply a proof that covers all cases.

It would have been truly great had the ancient Chinese been completely right in what they asserted. Why? Because it would have been a criterion for deciding whether a (large) natural number n is prime. Were it true, it would say, "Compute the residue 2^{n-1} mod n. If it comes out to be 1 then n is prime, if not then n is composite." Alas!

Important Remark: It is simple to compute b^{n-1} mod n for any natural number b using an algorithm known as the *fast exponentiation algorithm* (FEA), which is based on *repeated squaring*. We don't need to know how it works, only that it requires no more than minutes on an ordinary computer to compute b^{n-1} mod n for an n with several thousand digits.

So how did our nineteenth-century mystery person get that $341 | 2^{340} - 1$? I don't know, but perhaps (s)he noticed that

$$2^{10} = 1024 = 3 \times 341 + 1$$

so that $2^{10} - 1 = 3 \times 341$. This says that $341 | 2^{10} - 1$, or equivalently that

$$2^{10} \equiv 1 \pmod{341}$$

From this observation it follows, on raising both sides of this congruence to the power 34, that

$$2^{340} \equiv 1 \pmod{341}$$

Game, set and match. Thus $341 | 2^{340} - 1$. A pity.

Pseudoprimes

If it were the case that

$$2^{340} \equiv 5 \pmod{341}$$

then we would know immediately that 341 is a composite number.

Why again, exactly? Because if 341 were a prime, the FLT would assert that

$$2^{340} \equiv 1 \pmod{341}$$

Then $2^{340} \equiv 5 \pmod{341}$ would reveal 341 as a composite immediately.

However, the fact that $2^{340} \equiv 1 \pmod{341}$ does not in itself prove that 341 is prime. We say that the integer 341 passes the "FLT test for the base 2," but we cannot, as a result of this one test, say that 341 is prime. If 341 were prime, the FLT would also assert (because 3 is relatively prime to 341) that $3^{340} \equiv 1 \pmod{341}$.

This is where 341 is revealed to be composite. The fast exponentiation algorithm shows in a jiffy that

$$3^{340} \equiv 56 \pmod{341}$$

Now we know that 341 is composite. This result unmasks the composite nature of 341. The number 3 is said to be a *witness* to the compositeness of 341. It is worth thinking about what is taking place here.

A number (acting as modulus) which passes the FLT for a given integer a, but which is not a prime, is said to be a *pseudoprime to the base a*. Because 341 passes the FLT test for the number 2, the number 341 is a pseudoprime to the base 2. However, the number 341 is not a pseudoprime to the base 3. You might like to know that the composite number $91 = 7 \times 13$ is a pseudoprime to the base 3 because

$$3^{90} \equiv 1 \pmod{91}$$

Something very important and striking has just occurred. Because

$$3^{340} \equiv 56 \pmod{341}$$

we can say with certainty that 341 is a composite number. We are saying with certainty, *without* displaying a single (nontrivial) factor to support our claim, that 341 is composite.

If you are not impressed by that example, consider this one. Using the FEA it is simple to show quickly for the number 2, which is obviously relatively prime to 11,111, that

$$2^{11,110} \equiv 10,536 \pmod{11,111}$$

Thus we can say with complete confidence that 11,111 is a composite number. Simply amazing! And we haven't displayed a single factor. Can you find two nontrivial factors of 11,111? (Verify your answer in Appendix B, page 302.)

I realize that I have used the words "interesting" and "fascinating" many times up to this point (I checked—"interesting" eleven times and "fascinating" three times!). But nowhere have I used them with more justification than now.

Up until now, the only way we have seen of knowing that a number is composite was by displaying a (nontrivial) factor. But we have just seen another way. It is a major new insight into the nature of composite numbers to realize that the FLT can (sometimes) reveal the compositeness of a number n without exhibiting a single factor. To reiterate how it achieves this, if

$$a^{n-1} \not\equiv 1 \pmod{n}$$

for some a, then n is composite.

There are "large" numbers known to be composite in just this way, yet no one knows any of their nontrivial factors. When I first heard Dad mention this, it made me think how tremendously difficult factoring must be. Imagine knowing without doubt that a number is composite, yet being unable to produce a single nontrivial factor which would convince everybody of the number's compositeness by the familiar procedure of dividing the factor into the number. When I learned later that RSA Laboratories posted sets of these numbers on challenge lists, I saw it as more telling evidence that this factoring business is taken very seriously. I couldn't wait to hear the actual details of how different cryptosystems use the difficulty of factoring in a positive way.

The Parting of the Problems

Finding prime numbers and the inability to factor certain "large" numbers are essential to the security of many modern cryptosystems, facts that will, I hope, become clear in later chapters. Most people, and I was one of them, think that the problem of deciding whether or not a number is prime is one and the same as the problem of being able to factor numbers. As I now explain this is not so. Suppose we take an integer n and start "testing" it by calculating a^{n-1} for $a = 2, 3, 4, \ldots$, reducing modulo n, and suppose that on each occasion it happens that $a^{n-1} \equiv 1 \pmod{n}$. What can we say? Unfortunately, we cannot conclude that just because n has passed the FLT for this number of "bases," the integer n is prime. It may well be, and probably is, but we cannot be sure. Using this method we would need to test every base up to $a = n - 1$, and only then, if each base has passed the FLT test, could we with certainty pronounce n to be prime. This is clearly a totally inefficient way of testing that the integer n is prime, and is even worse than the trial division algorithm.

However, you'd be forgiven for thinking that if n is composite then some small base a will make $a^{n-1} \not\equiv 1 \pmod{n}$ and so reveal n as a composite very quickly. You might also be forgiven for thinking that if this does not happen for the first ten bases, say, then n is almost certainly a prime. Alas, neither is the case. There are composite numbers n which pass the FLT test for every base number relatively prime to n. They are called *Carmichael numbers** because they were discovered by the mathematician R. D. Carmichael in 1910. It has been known since 1994 that there is an infinite number of them, which is very sad for those of us who might have hoped to establish compositeness using the FLT test. The smallest of the Carmichael numbers is $n = 561$. Every a relatively prime to 561 makes $a^{n-1} \equiv 1 \pmod{n}$, though $561 = 3 \times 11 \times 17$ is composite. Here the base $a = 3$ would show that $a^{560} \not\equiv 1 \pmod{561}$ and so reveal this number to be composite with just two tests. However, not all Carmichael numbers have a small factor, so testing in this manner could prove very inefficient.

Nevertheless, there are refinements of the FLT test which declare very quickly whether Carmichael numbers are composite. If an integer n has not been shown to be composite after undergoing several such tests then it is classified as a probable prime, though it would not actually have been *proved* to be prime. These so-called Monte Carlo tests which produce probable primes are very fast and, while not providing certainty, can reduce the probability that the integer is not a prime to well below the probability of my winning the lottery two weeks running.

There are tests, called *deterministic primality tests,* which establish prime numbers to be so with certainty. They are slower than the tests that produce primes with a high degree of certainty, but many, known since the late 1980s, can provide proof of primality for integers in excess of 500 digits in a matter of minutes. This is an extremely important practical fact because, as we shall see, many modern cryptosystems rely fundamentally on the use of prime numbers of 100 digits or more!

Because of the FLT test and its refinements, we can determine quite quickly whether or not an integer of some hundreds of digits is a prime. The problem of factoring a known composite number with the same number of digits is considered much harder. It is crucial to understand that **deciding whether an integer n**

*Sometimes called absolute pseudoprimes.

is prime or composite is different from finding the factors of a random large composite integer. Because the trial division algorithm either tells that a number is prime or produces a factor of a composite integer, we are inclined to view both problems as being one and the same. It can take time to appreciate that testing for primality and providing a factorization are two separate problems, the first being relatively easy while the second is currently believed—though not proved—to be intractable.

The FLT depends vitally on the fact that the modulus p is a prime number. It is another striking example of the elite role that primes play among the natural numbers. If you try the FLT for a composite n, you'll find it won't work. For example, try $n = 8$ and $a = 3$. Does $8|3^7 - 1$? It doesn't, because $3^7 - 1 = 2187 - 1 = 2186$, which is not divisible by 8.

Leonhard Euler was the first to investigate the more general case where n is any natural number. He eventually found the answer in 1736. His result is slightly more technical than the FLT, but this is offset by the fact that it is more general, applying as it does to every natural number n. If you would like to know more about this important result, have a look at Appendix D, page 315.

Further Mathematical Excursions with Dad

The mathematics I have just described deals almost exclusively with whole numbers, or integers as they are termed. The area of study devoted to discovering their properties is renowned for being very difficult. Despite this, it has captivated the minds of some of the greatest mathematicians over the ages. There are numbers "beyond" the integers, the simplest and most familiar of which are the fractions. In case you have forgotten, a fraction has a numerator and a denominator. The numerator of the fraction $5/13$ is 5, while its denominator is 13. The fraction $5/13$ can be expressed in ratio form as 5:13. So, speaking a little loosely, you could say that fractions are expressible as the ratio of two whole numbers. For this reason they are often called *ratio*nal numbers, or rationals for short. Thus $1/2$ and

$^{23}/_4$ are rational numbers, though they are not integers. Integers, on the other hand, are rational numbers because they can be thought of as fractions whose denominators are 1.

For a long time it was believed that there were no types of numbers other than rationals. The ancient Pythagorean proclamation that "all is number" was an expression of this belief. But there are other species of number. Beyond the rationals are the real numbers, including such wonders as $\sqrt{2}$, the golden ratio ϕ, the base of the natural logarithms e and, of course, the ubiquitous π. These numbers, some of which, as you can see, have letters for their names, are considered "real" because they convince us of their existence by being geometrical ratios or other tangible quantities. There are also numbers which are not "real," known as complex numbers and built around the mysterious $\sqrt{-1}$.

Unfortunately all these exotic numbers play no part in the tale to be told so I must refrain from digressing for a couple of hundred pages to tell you "all" about them—not that I could. However, I cannot resist saying just a little about some of them.

As part of the Mathematical Excursions course, Dad gave a three-hour lecture (less twenty minutes for a tea break) on the four numbers $\sqrt{2}$, ϕ, e and π. I knew that π is an interesting number, I had never heard of e, and I thought I had heard about ϕ but I was mistaken. I fancied that there was nothing much to be said about $\sqrt{2}$ other than that it's that number which when squared gives 2. Of course, I should have known better! I was to learn that there is much, of an elementary nature, that can be said about $\sqrt{2}$. It was with its very definition that we started our excursion, teasing out what is really meant by it. We learned how upset the Greeks were when they first realized the "different" nature of this number—upset to the point, apparently, of drowning some poor devil just because he blabbed that all was not well with the accepted dogma that every positive number could be expressed as a ratio of two natural numbers.

The number $\sqrt{2}$ makes its presence felt by being the length of the diagonal of a square whose side measures one unit in length. Ironically, it is the famous Pythagorean theorem that tells us this.

So what is new or different about $\sqrt{2}$? Well, try as they might the Greeks couldn't find a rational number which when squared gave 2 exactly. No approximations, like $17/12$ (which incidentally doesn't do too bad a job), would do. No sir! It was to be all or nothing:

WANTED
a fraction whose square is 2.0000000000000000 . . .
with 0's all the way "to infinity."

They never found one. Being clever fellows, it gradually dawned on them that perhaps there is no such fraction, but how could they possibly prove such a thing? Well, I won't go on about it, but as you have probably guessed they did finally (about three hundred years after first suspecting that there was more to $\sqrt{2}$ than meets the eye) find an argument which showed beyond doubt that there is no fraction which when squared gives 2. So $\sqrt{2}$ was a new specimen.

The proof that $\sqrt{2}$ is irrational (that is, not expressible as a ratio of two whole numbers) so enthralled the English mathematician G. H. Hardy (1877–1947) as a boy that from the moment he read it, he decided to devote his life to mathematics. He did, and he became one of England's greatest mathematicians. He wrote a book called *A Mathematician's Apology*. It was Hardy who said, "Beauty is the first test. There is no permanent place in the world for ugly mathematics." The proof that $\sqrt{2}$ is irrational is another of those proofs by contradiction. It's a classic.

Dad showed us how the ancient Babylonians obtained a formula for finding rational approximations to $\sqrt{2}$ which is equivalent to one that was obtained later (after 1700) using a procedure known as Newton's method. He also showed how to derive the next term in the sequence

$$\frac{1}{1}, \frac{3}{2}, \frac{7}{5}, \frac{17}{12}, \frac{41}{29}, \frac{99}{70}, \ldots$$

The successive terms of this sequence give better and better fractional appromixations to $\sqrt{2}$. Try a few on a calculator and see. Just divide the numerator by the denominator and square the result. Have another look at the fractions above to see how you might get the next one from the previous one before reading on.

> To get the denominator of the next term in the sequence you simply add the denominator and numerator of the previous fraction, while to get the numerator you simply add twice the denominator of the previous fraction to its numerator.

Let's see how $7/5$ generates the next fraction according to this procedure. The new denominator is $5 + 7 = 12$, while the new numerator is $2 \times 5 + 7 = 17$. So the next fraction should be $17/12$. It is.

Dad showed us some consequences of the irrationality of $\sqrt{2}$ which I found startling but which gave me a deeper insight into what it meant to be irrational! The side of a square and its diagonal cannot both be measured exactly with the same ruler no matter how fine its markings. Can you believe that? A fancy way of putting this is to say that the side and diagonal of a square are *incommensurable*. If you have a ruler which measures the side exactly (meaning that the two end points of the side coincide with two markings of the ruler), then that same ruler when placed along the diagonal so that one of its markings coincides with the initial point of the diagonal will have the end point of the diagonal lying between two of its markings. Always. And if the ruler measures the diagonal exactly it will fail to measure the side exactly. If you had at your disposal a new ruler with ultra-close markings, or even an infinity of such rulers with every conceivable separation between the markings, it would make no difference. This geometrical fact gave me more respect for $\sqrt{2}$—it held some interest.

Dad gave a two-dimensional example of a rectangle in which the lengths of the longer and shorter sides are in the ratio $\sqrt{2}:1$. This rectangle cannot be filled with square tiles, no matter how small the tiles. We have all seen rectangles that are approximately this shape. If you look at a pad of A4 paper (the standard size used in the UK and Europe), you'll see that its cover says the pad measures 297 × 210 mm. Now, the last fraction in the sequence mentioned above is $99/70$. Multiply it "above and below" by 3 and you'll get $297/210$. The A4 sheet is supposed to have the length of its longer to its shorter side in the *exact* ratio $\sqrt{2}:1$ for a very practical reason. In this case (and in this case only, as a little algebra easily shows), the sheet can be folded along its longer side to give two smaller rectangles, for each of which the ratio of the longer to the shorter side is again $\sqrt{2}:1$. Therefore each of the new sheets thus formed has the same property as the parent. Two A4 sheets are the offspring of a single A3, and in turn, two A3s come from one A2. The first in the line of the A series is an A0 sheet whose dimensions are 1189 × 841 mm. This is approximately a square meter in area. Fold an A4 page in two and check that the ratio of the longer to the shorter side is approximately $\sqrt{2}:1$.

Of the other three numbers, ϕ, e and π, much was said about ϕ, but all that I remember now is something about a golden rectangle and the Fibonacci sequence. Most of what I heard about the number e has also receded into the memory's mists, but I have a faint recollection of a bank manager who was willing to pay 100 per cent interest per annum on an investment of £1. He was also willing to have interest added every six months at 50 per cent compound interest, or even to have interest compounded every day at the appropriate rate. He was even willing to be pushed to the limits of his resources and his generosity and allow the £1 to gather interest "continuously." I also remember that even with all this frantic adding of interest, the £1 does not exceed all known fortunes by the end of a year. Instead it becomes a modest £2.72. The decimal 2.72 is the number $e = 2.718281\ldots$ rounded to two decimal places. (By the way, the letter e is used in honor of Euler.)

As for our old school chum π, the latest news is that its decimal expansion is now known to 206,158,430,000 digits. Incredible!

We covered many interesting topics in the Mathematical Excursions evening class. One was floors and ceilings (nothing to do with building); another was magic squares and knights hopping all around the chessboard. (Speaking of which, I bet you think that there are 64 squares on the chessboard. Well, there are 204. How about that?) A further topic was quadratics and projectiles. In another session, a ball was dropping from far up in the sky and we were trying to figure out how fast it was going at any given instant—instantaneous speed. Calculus stuff.

But Dad was at his most dramatic when we discussed probability. This was when I realized that he was a closet gambler, if not a latent con man. During these talks, packs of cards, dice and other miscellaneous props such as toy cars and large mugs made their appearance. There was talk of winning lotteries, fellows getting on planes with bombs, birthday paradoxes, goats and cars behind screens, envelopes with sums of money in them, buying cornflakes boxes to collect models of Robin Hood and his Merry Men and, of course, playing poker.

Dad reveled in confounding our intuitions with bizarre examples. At one point he showed us three cards, one colored red on both sides, another colored blue on both sides and the other colored red on one side and blue on the other. "If you choose a card from one of these three and it's red on one side, what would you bet me that the other side is blue?" Don't be fooled! It's not a fifty-fifty bet (to use bookmakers' language).

We discussed decision-making based on assessments of probabilities, when to play and when to fold in poker games. We saw counter-intuitive card tricks, such as the one I just described, and of course we discussed the *Let's Make a Deal* game show with the aforementioned car and goats. This amazing problem appears in almost every modern popular book on mathematics and there are great stories connected with it, some of which are discussed at length in the book *The Man Who Loved Only Numbers,* a biography of the Hungarian mathematician Paul Erdös.

Now that I have almost surely whetted your appetite, it would be cruel of me not to state the problem. Here goes. A car is hidden behind one of three doors, and each of the other two doors conceals a goat. Lucky you, who gets to guess where the car is, choose door 1. Monty, the game show host, who knows where the car is, then opens one of the other two doors to reveal a goat. He asks you whether or not you wish to switch from your original choice of door to the other closed door. Should you switch? (For the all-important answer, along with a brief explanation, see Appendix B, page 302.)

I hope you'll be pleased to learn that, from here on in, you'll have very little to learn in the way of new mathematics. Thanks to all the hard work you've done, you're ready to meet some of the truly amazing ideas behind modern cryptography which I have been discussing.

8 One Way Only

Dad began one of our excursions into cryptography by saying that a fascinating aspect of the modern form of the science is that things that appear to be of a negative nature are exploited in a positive way. He hoped to illustrate this point by telling us about one-way functions and one-way trapdoor functions. The word "trapdoor" caught my attention immediately, and I wondered how such a term could be used in relation to functions. "One-way" conjured up an image of one-way streets. So already I was sure I was going to enjoy this lecture, though I was hoping it wouldn't get too technical. You know how it sometimes happens—you follow everything for the first ten minutes and then you wonder at what point the lecturer started talking double Dutch. Dad continued by saying that after he had described the nature of these "wonderful" functions we would probably ask him, "What possible use could anyone make of these?" In order to whet our appetites—and to ensure that we would listen attentively—he said, "Let me tell you about a little problem that Alice and Bob had which they solved using one-way functions."

Alice and Bob work for the same company, a very prestigious computer company. They are both managers at the same level, Alice in London and Bob in New York. The company has a very important upcoming conference in Tokyo, which they would both love to attend. Unfortunately, the company is willing to provide an all-expenses-paid trip for just one of them, and has left it up to them to decide who will go.

If Alice and Bob were in the same place they could decide the matter in the time-honored fashion by tossing a coin, but they are not. But the amazing thing is that they can simulate the fair toss of a coin. "How could they possibly do such a thing?" I hear you ask. Answer—by using a one-way function. The procedure they use is known as the remote coin-tossing protocol. When I first heard of this I was intrigued and captivated. Who gets to toss the coin, and how can the other person be sure that (s)he is not being cheated?

To describe to you how the two executives settled matters, I must begin by telling you about one-way functions. Strictly speaking, these are mathematical functions or transformations, but we can describe a one-way function loosely as an operation that is easy to do but extremely difficult, if not impossible, to undo.

For example, it is easy to mix a can of white paint with a can of black to obtain a mixture of gray, but none of us would relish the task of having to recover the original measures of white and black paint. It is easy to squeeze toothpaste out of a tube, but how would you get it to go back in? The good cup or dish that is all too easily broken into pieces is rarely recoverable. Humpty Dumpty was a victim of this element of irreversibility when he had his great fall off the wall, for

> All the King's horses and all the King's men
> Couldn't put Humpty together again.

Not all examples of one-way functions are so strongly suggestive of order descending into chaos without hope of redemption. If you have a person's name and address it is easy to look up his telephone number in a telephone book. Suppose, on the other hand, that you were at a party many weeks ago and had jotted down on a slip of paper someone's telephone number but have now forgotten that person's name. If you would dearly like to know who the person is, then you're in a pickle. Unless you are willing to risk giving offense by dialing the number without any idea of who it is you might be phoning, you may consider yourself as being on the reverse side of a one-way function. Whereas it is easy to find a telephone number corresponding to a name, it is a horse of a different color—given the construction of normal telephone directories—to find a name corresponding to a number. If

Eddie O'Hare

Dad and I, walking on the grounds of Blarney Castle.

Eddie O'Hare

Our home, with Blarney Castle in the background on the left. Cork lies beyond the hill.

Newspics

The family, in our conservatory, February 1999: Michael is between Mom and Dad with (left to right) Brian, Eamonn, and David in front of me.

The Flannery family

As a baby with Dad. I didn't grow any hair worth mentioning until I was two.

Newspics

With Mom, February 1999. I can't remember why I was looking so pensive.

The Flannery family

That's me at five and a half, with baby Brian, Michael, and Dad, in the summer of 1987.

Bryan Murphy

Jumping Clydie in Millstreet, County Cork, when I was seven ("Look up, Sarah, look up!"). My beloved Clydie had to be put down in 1999, when he should have been enjoying his retirement.

Sygma

At the blackboard, sketching out some of the ideas behind my cryptographic project.

Eddie O'Hare

Eamonn at the kitchen blackboard, working on a puzzle while Dad and David look on. As usual, Mom is the last to sit down.

The Flannery family

Discussing my project with Grandad at the Royal Dublin Society some hours after winning the 1999 Esat Telecom and Young Scientist & Technology Exhibition.

The Flannery family

Speaking with Ireland's President Mary McAleese at Áras an Uachtaráin (her official residence) on the Sunday after becoming Ireland's Young Scientist of the Year 1999.

The Flannery Family

In the computer room at school, January 1999. I'm holding my Young Scientist trophy in the company of (left to right) the school principal; a government official; my classmate Vincent Foley (holding his Intel Excellence Award); and my science teacher, Seán Foley, Vincent's father.

Eddie O'Hare

Working on a laptop in Dad's study during the writing of this book. Everything on the shelves relates to mathematics and cryptography.

Eddie O'Hare

Me again, pretty much as I look today.

you were desperate (and you would certainly need to be) you could work your way down the columns of numbers in what you hope is the appropriate directory until you found the number which would reveal the name you're after. The fact that it is not impossible, even if it is prohibitively time-consuming, to recover the name from the number is what makes this example of a one-way function more interesting and close to the kind of one-way function that cryptographers exploit. If you had a special directory which lists telephone numbers in numerical order and gives the corresponding names (these directories are not available to the general public, who would have little use for them), then the job of finding the name from the number would be no more difficult than the other way around. This special information, often glamorously referred to as "trapdoor" information, effectively undoes the one-way function. (More on this later when we . . .)

Here is another example. John and Heidi become the best of friends at summer camp. When Heidi returns to Germany, John, in order to impress his newfound friend, takes out his English–German dictionary and attempts to write a letter in German which he then sends to Heidi. A few days later a reply arrives from Heidi written in German. John, who knows only a little German, is unable to translate most of Heidi's words and is beside himself with curiosity as to their meaning. His own dictionary is not of much use to him; it is a one-way dictionary—English into German—and not, as he now needs, German into English. This situation is the same as the telephone directory with names corresponding to English words and numbers corresponding to German words. John has lots of numbers for which he desperately wants names. The only thing he can do is go out and buy a German–English dictionary, which, he is pleased to find, he can do. But it is not always so easy to find the trapdoor.

Let me give an example with numbers which illustrates clearly the key idea of "easy one way, hard the other way" that is central to the notion of a one-way function. (Don't worry, I haven't forgotten about Alice and Bob, but it will be awhile before I get back to them.)

Choose two prime numbers, p and q, say, each five or six digits long, from a book of primes and multiply them on your calculator to obtain their product, n. Give the calculator with the number n displayed on it to a friend. Tell him how n was formed and challenge him to use the calculator to find p and q. (Unique factorization guarantees that n can be recovered only by discovering p and q.) Don't hold your breath while waiting for the answer.

Here, the forward operation of *multiplying* is computationally easy, whereas the backward operation of *factoring* is computationally hard. You, who know how n was computed in the first instance, are in possession of the secret trapdoor information—the factorization of n into its factors, p and q.

Of course, if you allow your friend to use a computer equipped with a factorization algorithm then he will find p and q in a short while because the primes you chose are relatively small and current factorization techniques will be adequate to the task. But up the ante, and use the very same computer to choose two very large prime numbers, say of a hundred digits each (that this can be done is a story in itself), and form $n = p \times q$ as before. A computer can do this huge multiplication in the twinkling of an eye. Now show the product n to your friend and leave him with his super-duper computer to search for p and q. He will not find them.

Here is an actual 200-digit product of two primes, p and q:

31580479476026838343434028711320725847154882228947750
15428152228083683447368974676059807575462905948363955
56373144745242759035196176119637870933244742660580931
81460871296266561056469547820922800384429

Can you find p and q? Do you want to? Feel free to use all the computing power you can get your hands on. (Only consult Appendix B, page 303, as a last resort.)

Before I describe how the coin-tossing protocol works and return to Alice and Bob to see which of them is going to Tokyo, let me explain in simple mathematical terms what it means for a function to qualify as a one-way function. First and foremost, it must be a one-to-one function (meaning that there is a unique x corresponding to each y). If you have forgotten about the function notation and all this one-to-one business you could look back to Chapter 6, but really you can read on because you'll get the idea even if you are a little hazy about the notation. One warning, however: since the terms "one-to-one" and "one-way" are similar, be on your guard.

Here goes. A function f is a one-way function if

it is easy to compute $y = f(x)$ for each x

but

given y it is very difficult to compute $f^{-1}(y) = x$ (read as "f inverse of $y = x$")

All of which is in keeping with our earlier "easy one way, very hard the other way."

The function $f(x) = 2x + 1$, which is a one-to-one function, could hardly be described as a one-way function. It is easy to figure out from a given y the unique value of x that generates it. Using algebra,

$$y = 2x + 1 \Rightarrow 2x = y - 1 \Rightarrow x = \frac{y - 1}{2}$$

So this function holds no mystery and does not qualify as a one-way function. On the other hand, the cubing function

$$f(x) = x^3$$

is a little different. If you don't have a calculator or computer then my telling you that $y = 1{,}860{,}867$ and asking you for the value of x whose cube is this y might tax your computational abilities. It is not so easy to see how to unravel or invert the action of this cubing function. However, it can be done, and in this case you would find that $x = 123$. Although this function would not be regarded as a one-way function, we can see that there is much more to inverting it than there is with $f(x) = 2x + 1$ and we get an inkling of the nature of a one-way function. So to reiterate: a one-way function is one for which it is easy to compute $y = f(x)$, but not so easy, in fact downright hard, to figure out x given $y = f(x)$. If it can be done at all, it simply takes too much time. The phrase used is "computationally infeasible." Lots of one-way functions are "known," though if truth be told, nobody has ever given a strict mathematical definition of what is meant by "hard to invert," and so no one has ever proved that such functions really exist. A good lawyer might have a field day cross-examining us on what exactly a one-way function is. But, that said, mathematicians know from centuries of experience that certain mathematical operations are easy to do, but that nobody has ever figured out a smart and efficient way to undo them (in the same way that I can't prove to you there is no easy way to unmix those cans of white and black paint, but I'd be willing to bet you a lot of money that there isn't).

At last we can tell how Alice and Bob resolved their difficulty.

A Coin-Tossing Protocol

1. Alice and Bob agree on a one-way function f.
2. Alice chooses a random number x and computes $y = f(x)$.
3. Alice sends y to Bob.
4. Since Bob cannot compute x from a knowledge of y, he simply guesses whether x is even (heads, say) or odd (tails), and relays his guess to Alice. This is the "coin toss."
5. Alice tells Bob whether he won or lost the toss, and then sends x to Bob.
6. Bob confirms the result of the toss by verifying that $y = f(x)$.

The security of this protocol rests on the one-way function. Let us see what this is all about, step-by-step.

Step 1. How do Alice and Bob agree on the one-way function f? Well, we are assuming that everything is done by computers that have appropriate mathematical software. Alice has her computer on her desk in London, and Bob has his in New York. Furthermore, they have e-mail facilities so that they can communicate with each other. They are mathematically literate so they both know what a one-way function is and have access to any number of such functions. So to begin, Alice chooses a one-way function and sends it to Bob by e-mail for his approval. It is rather like saying, if they were together in the same room, "Are you happy that I toss this particular coin?"

Step 2. Alice chooses a random number x and computes $y = f(x)$. It should be said that the number of choices for x that Alice will have at her disposal will be truly enormous, but it is wise to get the machine to pick one such number from this myriad of possibilities using its random number generator rather than choosing herself. She may, without knowing it, betray certain biases such as favoring even numbers or ones ending in a 3.

Step 3. Alice calculates y by computing $f(x)$. She sends y to Bob but keeps knowledge of x to herself for the moment. This is all very straightforward.

Step 4. So now the number y pops up on Bob's computer screen in New York, and all he has to do is say whether or not the x that generated this y is an odd or even number. This is the coin toss: odd or even—heads or tails, a fifty-fifty bet. Clearly this is a crucial stage in the protocol. Bob knows the function f and also that $y = f(x)$ for some unknown integer x. Can he deduce what x is from these two facts? No! He has neither the time nor the ability to figure out from this information what x is because the function f is one of those nasty irreversible functions. This is where the one-way property of f is being set to work. There's nothing for it but simply to guess that x is even, say, relay the guess to Alice and hope for the best.

Step 5. Moments later, "Sorry, Bob! Better luck next time. The number x I chose is odd. I know you trust me, but here is my lucky x anyway."

Step 6. Of course Bob does not trust Alice, but because she has sent him an x he knows his goose is cooked. He has no doubt that when he applies the f, which he received at the outset from Alice, to x he will get the y that she sent him. This is how he verifies that there have been no shenanigans.

That is the gist of the protocol, which I found truly fascinating when I first heard it described. There are certain questions you might well ask. One is, "Could it be that Alice can have an odd x_{odd} and also an even x_{even}, each of which produces the y that she sends to Bob so if he guesses odd she swindles him by sending the x_{even} for verification?" No, because the functions used are one-to-one. Bob would know that in addition to being one-way, the function f sent by Alice is such that there is only one x corresponding to y. So that kind of skullduggery is ruled out.

However, there are subtleties and one needs to be vigilant. Suppose the cubing function were truly a one-way function (which it is not), and that Alice unwittingly asked for this to be the function f used as the one-way function for the coin toss. Bob would be secretly delighted and no doubt agree. When Alice sends him y he knows he cannot compute the x that produces this y, but he doesn't need to. He is simply being asked to say whether x is even or odd. What could you say about the oddness or otherwise of x if you were to receive a y that is even? (Check your answer in Appendix B, page 303.)

So Alice got to go to Tokyo. You didn't think I was going to let the man win, did you?

What I liked about the lecture on one-way functions and their exotic cousins, the one-way trapdoor functions, was that it was for the most part conceptual in nature. There was nothing more technical needed than an elementary knowledge of what is meant by a function and the notation employed, a little about one-to-one-ness and an understanding of a one-way function. It was really about the power of simple ideas that are imaginative and easily grasped. When I said this to Dad he was delighted, telling me that he had read somewhere that the real moving force in mathematics is imagination, that it was a subject full of truly imaginative ideas. He said that David Hilbert (whom we mentioned earlier), on being told that some student had forsaken mathematics for poetry, quipped, "It is just as well, he didn't have enough imagination to be a mathematician."

Passwords

As children we learn how a certain word or phrase is used as a way to identify ourselves so that we can *pass* into a place to which access would otherwise be denied. We understand that the purpose of such a password or pass-phrase is to prevent unauthorized persons from gaining admittance to places where they are not supposed to be and where they would, more than likely, get up to no good. Every child has gazed with rapt attention at the TV or movie screen to hear if the hero can answer to guards' "Who goes there?" with something more than "John Smith." The good guy must be able to convince the guards that he is indeed the John Smith he is pretending to be, by reciting the prearranged word or phrase that the real John Smith would know. If he fails to give this vital password then the very least he can expect is to be denied entry.

Because most computers cannot yet "see," they can't recognize humans by sight. Thus when you wish to gain authorized access to a particular computer (or the section assigned for your own private use) you must convince it that you are who you claim to be. The normal way of doing this is to tell the computer who you are the first time you use it, and give it a password or pass-phrase known only to you. The computer keeps a record of this information for future use. The next time you wish to use this same computer, you enter your name and it requests your password. When you supply the password the computer checks it against the logged version and grants you access only if it matches. When this happens you have successfully "logged on." If someone knows your name and wants to impersonate you to gain access to the information you have on the computer, he'll be foiled during this log-on procedure when he cannot supply the password: he'll be denied access because you have "password protected" your computer. This same principle is used to protect an e-mail account (usually with a different password) or access to an ATM machine, this time with a PIN (personal identification number) rather than a password.

This is a good system, but there is always the threat of someone using ingenious means to learn your password and the passwords of others. One way for an evil-minded user of the same computer system to do this—and it has been done—is to gain access to the file where *all* the passwords are stored. This is where one-way functions play a key role in protecting such a password file. They protect the passwords entered by each user of the system by enciphering them and storing only the enciphered version. This allows a password to be kept absolutely private so that not even the system administrator knows it. When a "word" is first submitted as a password, the system encrypts it with the one-way function, and places the resulting "string" of gobbledygook in a file which is also called a password file. No record of the word submitted is kept. When the user logs on at future times and enters the password when requested, the same one-way function is applied to it and so produces the same ciphertext string as before. The (encrypted) password file is then searched for a matching string, and when it is found access is granted to the system. A very good scheme.

Part of the great story Clifford Stoll tells in his book *The Cuckoo's Egg* is an account of how a hacker "got around" a one-way function and netted a plaintext password. Stoll had been tracking this hacker (called H from now on) for months without H's knowing it. (Read the book—I couldn't put it down!) One night Stoll saw H downloading a password file from a certain Unix machine. (H had figured out how to get into this machine some time previously.) "Why is H doing this?" Stoll asked himself, wondering what possible good could it do for H to be copying down all this gobbledygook when, surely, H knew that the one-way function guarantees that the strings cannot be decrypted to reveal the corresponding English passwords. So what was afoot?

Three weeks later Stoll saw H back in the same computer system, attempting to log on as a private user. Stoll waited and watched. A password was requested. One was duly entered and H was "in"—into someone's private account! What happened? Rest assured that H did not find this password by finding the inverse of the one-way function that would have allowed the password file to be deciphered. Answer: H had a Unix machine that used exactly the same one-way function to encipher passwords. So the clever but devious H, with possibly a lot of help from collaborators, had painstakingly submitted every single word from an English dictionary to this machine as a password and created a file of encrypted passwords, keeping track of the strings generated by each English word. This done, H compared the entries in this file with the one downloaded weeks earlier, hoping for a match. The extraordinary pains H had taken paid off. A match was found. H had a (plaintext) password.

Because Stoll alerted the computer world to this possibility, it is now recommended that a password should not be an English word on its own. Should you wish to use some easily remembered word, such as the name of a boat, as was used by one person mentioned in Stoll's book, then use it as part of a longer string of plaintext characters containing numbers and other symbols, preferably interleaved with the English characters. This simple expedient makes the above "attack" much, much harder. Had this been done by each user at the computer installation penetrated by H, each password would have remained beyond reach.

Where's the Trapdoor?

Now that I knew the idea behind a one-way function I was curious to know what a one-way trapdoor function might be. That extra term, "trapdoor," between one-way and function was intriguing. Before telling us, Dad did "the teacher bit" and had us remind him what a one-way function is. So to get things moving, somebody said, "Something which is easy to compute one way but virtually impossible the other way." Well, words to that effect.

By way of response Dad said, "Put slightly more technically, if f is a one-way function then it is easy to work out $f(x)$ to get a number y, but if only y is known then it is computationally infeasible to find x."

We know. Forget about trying unless you have a few thousand computers and a lot of time because there isn't a realistic hope of finding x from y. Come on, tell us what a one-way trapdoor function is.

"A one-way trapdoor function is a function which is not a one-way function."

He's not serious, surely? He's expressing himself this way for effect!

"You can't be saying that 'trapdoor' is just a very fancy way of saying 'not.' "

"In a sense, yes. A one-way trapdoor function is *not* a one-way function for the person who dreams it up. But for the rest of us, it is still a one-way function and this is all-important."

Curiouser and Curiouser!

He continued, "The designer constructs the function in such a way that only he knows how to find x from y, only he knows how to invert the function. This special information, which he keeps to himself, is called the trapdoor information. Apparently this term was inspired by those two-clown stage acts in which one clown pretended to the audience that only he knew where a trapdoor was in the stage floor, while his colleague didn't. When the one ignorant of its existence stood over it, the clown (with the trapdoor information) would have it spring open, and the unfortunate one would disappear down through the floor to howls of laughter."

We have all seen films where the hero, who is being chased by an evil bunch out to capture or kill him, dashes into a library with no exit other than the door he just entered. While we are certain that he is to be caught by his pursuers, who are hot on his heels, he makes towards a book behind which is a mechanism which causes a bookcase to open. He slips quickly through the opening and escapes along a secret passage while the bookcase swings back to its original position. When the villains enter the library moments later, we witness their astonishment at finding no trace of him. Even if one of them is quick-witted enough to guess what must have taken place, we and they know that there is no hope of finding that special book (which is the hero's trapdoor information) in time to continue the chase.

An American Intellectual Revolution

In 1976, Martin Hellman and Whitfield Diffie had the imagination and foresight to see how one-way trapdoor functions could revolutionize cryptography. They outlined their wonderful ideas in a paper entitled "New Directions in Cryptography," and acknowledged independent ideas of Ralph Merkle which appear in his paper "Secure Communications over Insecure Channels." So, without further ado, let me tell you something about this new cryptography.

Alice, Bob, Claire, Denis and others each design a one-way trapdoor function. Call the one that Alice designs f_A (read as f sub A). Similarly, let f_B be Bob's one-way trapdoor function, and f_C and f_D the respective one-way trapdoor functions of Claire and Denis. This done, they submit these functions to a public directory along with their names and other relevant information. The table

Public Directory	
Alice	f_A
Bob	f_B
Claire	f_C
Denis	f_D
⋮	⋮

shows names (only) along with the corresponding functions—on display for all to see, just like names and numbers in a telephone directory. Except to the person who submitted it, any one of these is a one-way function. This means that no one can figure out in a realistic amount of time, even with unlimited resources, how to invert it—no one, that is, other than the person who submitted it to the directory. (S)he (presumably) already knows how to invert this function because (s)he possesses the special trapdoor information that accomplishes this.

Alice, Bob, Claire, Denis and company keep secret the inverse function corresponding to the function which each has made public. Alice keeps knowledge of f_A^{-1} (read "*f* inverse sub A") entirely to herself, as does Bob with f_B^{-1} and Claire and Denis with f_C^{-1} and f_D^{-1}. Anyone submitting an *f* to this public directory must keep secret the f^{-1} that undoes the action of *f*.

So? Well, these public one-way functions can now be used as enciphering transformations to send messages in encrypted form between any or all of the subscribers to this public directory.

For example, if Alice wishes to communicate with Bob then all she need do is look up Bob's enciphering transformation, f_B, in the public directory. When she types her message on a computer screen she applies f_B to it and sends it by e-mail to Bob. On receipt of this enciphered e-mail he applies the deciphering transformation f_B^{-1} (which is kept secret on his computer) to decipher the ciphertext and so read Alice's message. Anyone who eavesdrops on this message cannot decipher and read it even if (s)he knows it is intended for Bob.

An important point: Alice doesn't have to know Bob to be able to write to him in secret. This is the same as my looking up the telephone number of someone I don't know personally and phoning him.

9 Public Key Cryptography

Strictly speaking, it is not the enciphering functions f_A, f_B, f_C and f_D which Alice, Bob, Claire and Denis submit to the public directory, but the keys associated with these functions. By the same token, what they keep secret are the corresponding deciphering keys. In general, each uses the same type of mathematical function but with an individual enciphering key. Each must keep the corresponding deciphering key absolutely top secret.

This form of cryptography, based as it is on a public directory which contains the enciphering key of each user, is known as *public key cryptography*, and any particular cryptosystem which implements such a scheme is called a *public key system*, or PKS for short.

So why is public key cryptography such a big deal—"revolutionary" as I described it earlier?

First of all, it is actually shocking to the point of being almost unbelievable that you can tell the whole world your enciphering key *without* this information revealing anything about your deciphering key. A public key system is based on publishing, for the whole world to see (legitimate users and others) the means by which messages are enciphered—in the full confidence that this knowledge alone does not lead to the deciphering processes being discovered in a realistic amount of time. Before the advent of public key cryptography, cryptographic systems (which we now call private key or classical systems) depended for their security on both keys being kept private. With such systems, if one of the keys becomes known, it is computationally feasible to obtain the other one. (In the Caesar system, knowing the decryption key reveals the encryption key immediately.)

What is revolutionary and enormously practical about public key cryptography is that it saves an awful lot of hassle! Why? Because it provides a solution to what is known as the problem of *key distribution*. Since it is necessary to keep both keys in a classical system private, Alice and Bob must first of all meet in secret if they want to fix on a classical system and on the keys to be used. (If they don't wish to do this, they must get a trusted courier to act as a go-between.) This is not always convenient, particularly if one lives in Dublin and the other in Berlin. They cannot trust e-mail or any other insecure form of communication. Now take a million (which is not a lot in the modern world of business) such Alices and Bobs, all wishing to talk to each other in private. How many such meetings must there be? (For 10 people, 45 such meetings are needed. See Appendix B, page 303, to confirm your guess as to why.) Answer: about 500,000,000,000, or half a trillion. That's a lot of secret meetings, necessitating a lot of travel and time.

With public key cryptography, there is no need for any of this globetrotting. Just look up the public directory to get the other person's enciphering transformation, and write to that person in private to introduce yourself.

Another revolutionary aspect of certain public key systems is the really smart way they solve the problem of *message authentication* while achieving a little more besides.

A Signature Scheme

Bob, who is a technologically very up-to-date bank official in Very-Rich Bank, received the following ciphertext:

$3iu%1bdfe&nKr7qn\$khg&*)vd\$1\}rAbz9hcq21bv%8&£22'/
Bd15sdvso£*sa6ph&gpsenbu*&%ks\$8g%g8s\$!kb8kkY
Tu11arq8(ytifan£19dr5 h\$ghkt6nvt&93sjlH

on his computer terminal on December 12, 2000.

When he applied his private deciphering transformation f_B^{-1} to the ciphertext, he obtained the following interesting message:

Dear Bob,

Would you please transfer one thousand pounds from my account to my sister Joan's? I want to give her a surprise Christmas present.

Merry Christmas!

Alice

How can Bob be sure that it really was Alice who sent this very nice message?

Since anyone wishing to impersonate Alice can look up Bob's enciphering transformation f_B in the public key directory, might it not be the case (perish the thought!) that Joan actually sent this message that purports to come from her generous sister Alice? How can a message such as this be authenticated?

If the public one-way trapdoor function f has the additional property that

$$f\,[f^{-1}(P)] = P \text{ (read as } f \text{ acting on } f \text{ inverse of } P\ldots)$$

then Alice can append a signature S to her message, which allows Bob to authenticate it.

Before I describe how, note that the property just described means that the deciphering transformation is first applied to a piece of plaintext (to give gobbledygook, as it will) and then the enciphering transformation is applied to recover the original plaintext. This is the opposite of what normally goes on in the enciphering/deciphering process. (In the Caesar system it would mean moving back three characters first and then forward three.)

Alice does the following:

1. She composes a "signature," S. (I'm being deliberately vague about its nature.)
2. Alice then applies her deciphering transformation f_A^{-1} to the signature S (she is the only one who knows this function f_A^{-1}) and appends the result, $f_A^{-1}(S)$, to the end of her message. (Often it is this $f_A^{-1}(S)$ that is thought of as the signature.)
3. She then enciphers both her message and her "decrypted" signature, $f_A^{-1}(S)$, using Bob's enciphering transformation f_B, and sends the encrypted document to Bob.
4. Bob deciphers the incoming ciphertext to learn that the message purports to come from Alice. When he applies f_B^{-1} to the signature portion $f_B[f_A^{-1}(S)]$ to get $f_A^{-1}(S)$, it appears as gobbledygook. He now looks up Alice's enciphering transformation f_A in the public directory and applies it to $f_A^{-1}(S)$ to get

$$f_A[f_A^{-1}(S)] = S$$

In this way the signature is recovered and the message is authenticated.

Is this good enough? It is on the face of it, because there seems to be no doubt that S could have come from anyone other than Alice. Nobody, but nobody, except Alice, knows f_A^{-1}.

What if Alice is careless and Bob is a baddie? Suppose that Alice's signature consisted of the single word "Alice." What could Bad Bob do? Almost anything, because he could cut and paste this decrypted signature onto other messages written by him but purporting to come from Alice. If he were cheeky and crazy enough he might even instruct his own bank to transfer funds from Alice's account to his. Should there be any raised eyebrows at all about the money now on its way to his account, he can no doubt point to a gushing letter extolling his virtues as the friendliest bank teller alive, and if this glowing account does not dispel all doubts about the transaction then he need only point to the signature, which could only have come from Alice.

So, ingenious as this signature scheme is, it needs a little fine-tuning to be used properly. This is achieved by making the signature S, in addition to being *signer*-dependent, *message*-dependent. Otherwise the message could be modified. Had Alice composed a signature S along the lines

> Re transfer of £1000 to Joan's a/c, Alice, 10.00 a.m., 12/12/00

containing as it does her name, a brief résumé of the message, and the time and date, then Bad Bob could do nothing.

The signature scheme I've described does a little more than allow a message to be authenticated. It also has the feature that the message cannot be repudiated. Should Alice have a change of heart at a later time, she cannot deny having sent this message since nobody but she knows f_A^{-1}.

What could Alice do if in the above signature she failed to specify the exact amount of money, and Bad Bob worked his charms on Joan to collude with him in increasing the amount by a factor of ten or even a hundred? You can never be too careful!

The RSA Public Key Cryptosystem

Although Diffie and Hellman introduced the concept of one-way trapdoor functions and explained to the world the radical consequences of their existence, it was Ronald Rivest, Adi Shamir and Leonard Adleman of MIT who, in 1977, first exhibited an example of one which is fairly simple and lends itself to easy implementation. No one has ever proved a purported one-way function to be one, but the three inventors of what is now known as the RSA cryptosystem had very strong empirical evidence that the function they employed is indeed one-way in character. The scheme they developed is based in part on the belief that it is tremendously difficult to factor "large" composite integers, a belief backed by centuries of failed attempts to find one single efficient method.

I shall describe this celebrated system, which has withstood all forms of attack since it first saw the light of day in the August 1977 issue of *Scientific American*. Although it is unavoidably technical, it is relatively simple to describe, and anyone who has followed the mathematical exposition this far, even superficially, will be able to appreciate the general character of the scheme and see most of the mathematical ideas coming into play. Just read on and don't worry. You'll be surprised and rewarded by obtaining some notion of what is involved.

Start-up: This need be done only once.

1. Generate, at random, two prime numbers, p and q, of 100 digits or more.
2. Calculate $n = p \times q$ and $\phi(n) = n - (p + q) + 1$.
3. Generate, at random, a number $e < \phi(n)$ such that e is relatively prime to $\phi(n)$.
4. Calculate the multiplicative inverse, d, of e modulo $\phi(n)$, using the Euclidean algorithm.

Publish: Make public the enciphering key,

$$K_{\mathrm{E}} = \langle n, e \rangle$$

Keep Secret: Conceal the deciphering key,

$$K_{\mathrm{D}} = \langle n, d \rangle$$

Enciphering: The enciphering transformation is

$$C = f(P) = P^e \bmod n$$

Deciphering: The deciphering transformation is

$$P = f^{-1}(C) = C^d \bmod n$$

That's all there is to it!

"OK, but what is it all about, and what does $\phi(n)$ mean?"

I'll say something about $\phi(n)$ a little later (I also discuss it in Appendix D, page 315), but there is no need to get distracted by it right now. Just think of it for the time being as a (certain important) number.

Suppose the public key $K_E = \langle n, e \rangle$ published above belongs to Bob. Such a key might look like

31580479476026838343434028711320725847154882228947750
15428152228083683447368974676059807575462905948363955
56373144745242759035196176119637870933244742660580931
81460871296266561056469547820922800384429

56373143474524275903751961761143434028711320725847154
88274266058093181460871298047947602626897467605983324
41742151212121810181381460871296326467596812035160486
6935241758428079727723

—two numbers which are pretty awesome in size; the first is Bob's public modulus n and the second is his public encryption exponent e. These are all I need to know to be able to send a message to Bob in encrypted form.

When I have written my message I divide it into blocks, called message units, and assign each a number P (the numerical equivalent of the message unit). I must choose the block size so that no P ever exceeds n, but this is easy to do. Now I encipher each P by calculating

$$C = P^e \bmod n$$

in turn to get a stream of C-numbers (the corresponding ciphertext numerical equivalents). *This is where I am making use of Bob's public encryption key,* $K_E = \langle n, e \rangle$. Of course, I have a program (written or purchased) which does all I have just described automatically in a matter of seconds.

This done, I press the "Send" button, and a file with a seemingly meaningless mess of gigantic numbers makes its way to Bob's computer. When Bob receives this file of C-numbers he takes each in turn and calculates C^d mod n. *This is where Bob makes use of his private or secret decryption key,* $K_D = \langle n, d \rangle$. (It is only the d-number that is secret, as n is known.) Amazingly, he gets back the original set of P-numbers, which he then converts to plaintext so as to read my message. Of course, like me he has a program that carries out all these steps automatically. In fact, the program is pretty much the same as the one I use because the encryption and decryption procedures in the RSA are exactly the same in character—I calculate P^e mod n, while Bob must calculate C^d mod n.

It is fascinating that in order to recover the P-numbers that I disguised as C-numbers by raising them to the power e, Bob actually reraises! Rather than extracting eth roots by some complicated algorithm, he succeeds in undoing the first power raising by raising the C-numbers to the power d.

This "poker tactic," as Dad calls it, works because e and d have been chosen very cleverly and it is the Euler Fermat theorem of 1736—or, if you want to stretch a point, Fermat's Little Theorem of 1640—which explains how and why it works. I'll spare you the details (which are not that difficult) and say only that these number theory results are pivotal in establishing the soundness of the RSA encryption/decryption procedure. Instead, I present an example with an unrealistically small modulus which illustrates the procedure. It is fun just to see a concrete example.

Start-up: The first thing we must do is generate the two primes, p and q, at random. Since we are keeping the numbers "small" for the purposes of this example, let us imagine that we look up a table that lists all the primes less than 1,000 and chose p to be 281 and q to be 167.

Then we calculate

$$n = p \times q = 281 \times 167 = 46{,}927$$

and

$$\phi(n) = n - (p + q) + 1 = 46{,}927 - (281 + 167) + 1 = 46{,}480$$

The number n we will publish as the public modulus. Knowing that $\phi(n) = 46{,}480$ allows us to choose the enciphering exponent e.

Our third task is to choose at random a number less than $\phi(n) = 46{,}480$ and check (using the Euclidean algorithm) that it has no factor other than 1 in common with 46,480. Let us suppose that on this occasion the first number we find meeting this criterion is 39,423. Thus $e = 39{,}423$ is the public exponent.

Finally, we use the Euclidean algorithm again, to calculate that the multiplicative inverse of e modulo $\phi(n)$ is given by $d = 26{,}767$. This is the vital deciphering exponent that is kept secret.

This completes the start-up procedure.

Publish: $K_E = (46{,}927, 39{,}423)$ which is our public key and keep secret $K_D = (46{,}927, 26{,}767)$, our private key.

This means that C-numbers are calculated from P-numbers by the rule

$$C = P^{39{,}423} \bmod 46{,}927$$

while P-numbers are recovered from C-numbers by the rule

$$P = C^{26{,}767} \bmod 46{,}927$$

However, only we know this all-important deciphering rule since we are the only ones who know the deciphering exponent d. Thus $C = P^{39{,}423} \bmod 46{,}927$ is our one-way function.

Enciphering: Let us suppose that the plaintext uses only the 26 characters of the lower-case alphabet. Since $26^3 = 17{,}576$ is less than the public modulus $n = 46{,}927$ while $26^4 = 456{,}976$ is greater, the plaintext may be broken into trigraphs (blocks of size 3) for enciphering. However, since the enciphering rule may produce C-numbers anywhere between 0 and 46,926, the ciphertext must be broken into blocks of size 4 to accommodate numbers in excess of 17,576 and less than 46,927.

Thus, to illustrate the enciphering and deciphering procedures we need only show them in action, first enciphering a single trigraph into a ciphertext unit of block-size 4 and secondly deciphering this ciphertext block to reproduce the original plaintext message unit. Let us choose the plaintext word "yes," and imagine it to be the top-secret reply to a plaintext message that said **"shall I sell all shares immediately?"**

OK. The first thing someone who wants to write to us in private does (in reality, the RSA program he has does) is assign the trigraph **yes** its numerical equivalent. This can be done in the manner described in Chapter 6. Since **y** is associated with the number **24,** **e** with **4** and **s** with **18,** the numerical equivalent, or P-number, assigned to **yes** is $P = 16{,}346$ since

$$\mathbf{24}.26^2 + \mathbf{4}.26 + \mathbf{18} = 16{,}346$$

Then the program uses the fast exponentiation algorithm to raise this number to the public exponent $e = 39{,}423$ and reduce it modulo the public modulus $n = 46{,}927$ to find that

$$C = 16{,}346^{39{,}423} \bmod 46{,}927 = 21{,}166$$

This number is converted in a 4-digit base-26 number to determine the ciphertext unit. Since

$$21{,}166 = \mathbf{1}.26^3 + \mathbf{5}.26^2 + \mathbf{8}.26 + \mathbf{2}$$

and **1** is associated with **b,** **5** with **f,** **8** with **i** and **2** with **c,** the ciphertext block **"bfic"** is the encipherment of the plaintext block **"yes."**

In summary, using the public key $K_E = (46{,}927,\ 39{,}423)$,

$$\textbf{yes} \rightarrow \textbf{bfic}$$

Deciphering: When our computer receives **bfic** our RSA program reassigns this 4 block its numerical equivalent, 21,166. Then, using the fast exponentiation algorithm, it raises this number to the power of the private exponent $d = 26{,}767$ and reduces it modulo 46,927 to get

$$21{,}166^{26{,}767} \bmod 46{,}927 = 16{,}346$$

This, then, is the numerical equivalent of the plaintext trigraph, which we write as $\mathbf{24.26^2 + 4.26 + 18.}$ Since the numbers {**24,4,18**} correspond to {**y,e,s**},

$$\text{yes} \leftarrow \text{bfic}$$

and we are done.

YOU: But doesn't the enciphering/deciphering procedure take an eternity, considering the enormity of the numbers involved? Any P or C can be hundreds of digits in length, and these numbers must be raised to the exponents e and d, respectively, each of which also can be hundreds of digits in length!

ME: Yes.

YOU: Well, won't the resulting numbers go through the roof?

ME: No.

YOU: Why not?

ME: Because all the arithmetic is done with respect to the modulus n, which means that all calculations are done on numbers which never exceed n.

YOU: I suppose that is something, but it must still take quite some time since n is still an enormous number.

ME: Computing $P^e \bmod n$ using the fast exponentiation algorithm requires about 600 calculations when e is a 100-digit number. More efficient procedures are known.

YOU: Hmm . . .

The *exponentiation* steps—that is, the calculation of $P^e \bmod n$

or C^d mod n in the RSA—are what make it, unfortunately, slower than classical cryptosystems. For this reason it is normally used on a one-time basis to decide on a classical system. The RSA algorithm works fine on short files, so it is ideal for deciding on—and exchanging—the keys of a classical system, which can then be used in all subsequent communications. At the moment, public key algorithms are about 1500 times slower than the classical ones available. But who knows? Somebody might come up with a system that will give the classical systems a run for their money. There is no suggestion that classical systems are less secure than public key systems, but the latter have some very attractive features. We live in hope.

Any More Questions?

There are millions of questions we can ask in connection with the RSA. Some are

1. Does it work?
2. Is it fast?
3. Is it secure?
4. What is the one-way trapdoor function?
5. Where is the trapdoor?
6. What is the current state of factoring?
7. Does the start-up procedure take long?
8. Are there enough primes?
9. Might two users pick the same two primes?

Let's take these questions in turn. As I've said, it does work (1) even if it is not as fast as one might like it to be (2). Is it secure? R, S and A in their original paper remarked that "since no techniques exist to prove that an encryption scheme is secure, the only test available is to see whether anyone can think of a way to break it." Well, since 1977 some of the world's finest mathematicians (who could call on any amount of computing power) have failed to find an attack that proves fatal to the RSA system. As its originators said in the same paper, "once the method has withstood all attacks for a sufficient length of time it may be used with a reasonable amount of confidence." That time is considered to have come, since the RSA is now accepted as a worldwide standard (3).

The one-way trapdoor function in the RSA is $f(x) = x^e \bmod n$. Apparently no one knows an efficient way of inverting this function without being able to factor the modulus n (4). Since knowing the number d allows the function to be inverted with ease, d is regarded as the trapdoor (5).

The status of factorization in 1997 can be seen from the following table, which gives the estimated time it would take to factor a number of so many decimal digits using the best-known algorithms available at the time (based on the assumption that the most basic operation takes a millionth of a second):

Number of decimal digits	Time (years)
150	<200
230	100,000
300	3×10^7
400	3×10^9

To help put the last two of these figures in perspective, the age of our planet is about 5×10^9 years (6).

YOU: You promised you'd explain what $\phi(n)$ means.

ME: I did. It stands for the number of numbers less than n which are relatively prime to it.

YOU: So $\phi(26) = 12$ because, as you told us some time ago, the numbers less than 26 which have no factor other than 1 in common with it are the twelve numbers 1, 3, 5, 7, 9, 11, 15, 17, 19, 21, 23 and 25.

ME: Exactly. It is very easy to say what value $\phi(n)$ has when n is a prime p.

YOU: Let me think. All the numbers less than p have no factor other than 1 in common with it because the only factors of a prime p are 1 and p.

ME: Yes.

YOU: So, $\phi(p) = p - 1$ since $p - 1$ is the number of numbers less than p.

ME: Top class!

YOU: Is it difficult to explain why $\phi(n) = n - (p + q) + 1$, as stated in the start-up procedure?

ME: Not really, but are you sure you want me to?

YOU: Why not? It can't be that hard, the answer looks simple enough. You just subtract p and q from n and add back 1.

ME: Right. The only numbers which have a nontrivial factor in common with $n = p \times q$ are either multiples of p or of q, since p and q are primes. If $p = 5$ and $q = 7$, so that $n = p \times q = 5 \times 7 = 35$, then the multiples of 5, namely 5, 10, 15, 20, 25, 30 and 35, along with the multiples of 7, which are 7, 14, 21, 28 and 35, are the only numbers with a nontrivial factor in common with 35.

YOU: I'm with you so far.

ME: Now, notice that the set of multiples of p contains q numbers, while the set of multiples of q contains p numbers. Check the example to see.

YOU: OK. There are seven multiples of 5 and five multiples of 7. Great!

ME: Now, these two sets of multiples have no element in common, other . . .

YOU: . . . than $p + q$, so there are $(p + q) - 1$ numbers with a nontrivial factor in common with n.

ME: Yes.

YOU: I've got it now. There are $n - [(p + q) - 1] = n - (p + q) + 1$ numbers relatively prime to n. That's why $\phi(n) = n - (p + q) + 1$. I'm a genius!

ME: I've never doubted it.

YOU: Can we go back to the RSA?

ME: By all means.

YOU: Might you not be able to get at d, the multiplicative inverse of e modulo $\phi(n)$, by searching?

ME: It is possible—but not feasible—if Bob goes about his business properly.

YOU: All right, then, calculate $\phi(n)$ and you're within reach of d because Bob has revealed its inverse, e. Since d is the inverse of e, just calculate e's inverse to get d.

ME: Absolutely right on, but how did you calculate $\phi(n)$?

YOU: Hmm, let me see. Well, I know n, so couldn't I just calculate all the numbers relatively prime to n?

ME: And how would you do that?

YOU: Oh! I'd have to test every number less than n with Euclid's algorithm. That's out of the question because n is so big. If this approach made it possible to do the calculation quickly, I'd actually end up factoring n.

ME: Correct.

YOU: Is there some other way of finding $\phi(n)$ without knowing the p and the q which are the factors of n?

ME: Nobody has found a way, and if they did it would lead to a factorization of n straightaway.

YOU: Interesting. I'll think about this, but of course it does mean that finding $\phi(n)$ must be as hard as the factoring problem.

ME: Yes. It is of paramount importance that $\phi(n)$ cannot be calculated, for a large $n = p \times q$, without knowledge of p and q.

YOU: So, you've got to know your p's and q's.

ME: Apparently.

The start-up procedure does not take long. Once the primes p and q are generated, it is only a matter of microseconds to compute n and $\phi(n)$, and a matter of seconds to determine e and its multiplicative inverse, d, modulo $\phi(n)$ using the Euclidean algorithm. The fascinating part, which we have already discussed in connection with Fermat's Little Theorem, is that one can find 100-digit primes in a matter of minutes (7). Furthermore, there is absolutely no shortage of primes of this length because the prime number theorem can be used to show that there are approximately 4×10^{97} of them (8). One procedure uses a device which generates 100-digit numbers (of which there are exactly 9×10^{99}) at random; each subsequent odd number is then tested for primality until one which is prime is found. If you generate two primes in this way, the chances that someone else has one or other of them as part of their public modulus are less than my chances of winning the lottery ten weeks in a row (9).

One final point on the RSA system: Its one-way trapdoor function has that all-important property necessary for a signature scheme. If a signature S is first raised to the power d modulo n and the result raised to the power e modulo n, then the original S reappears.

Official Secrets Out

In Simon Singh's *The Code Book* there is a section headed "The Alternative History of Public-Key Cryptography" which tells a story I find a little sad. An Englishman named James Ellis, working for the Communications-Electronics Security Group (CESG) at the Government Communications Headquarters (GCHQ) in Cheltenham, came up with the concepts of public key cryptography before 1970. Thus, he anticipated Diffie and Hellman by about six years, but the world was not to know of his ideas at the time because he was sworn to secrecy. As Ellis himself relates, he was inspired by a Second World War Bell Telephone report by an unknown author who described an ingenious idea for secure telephone speech. Ellis referred to his form of cryptography as "non-secret encryption" and, like Diffie and Hellman, he had no definite scheme worked out which would actually implement the ideas.

About six weeks after joining the CESG in September 1973, a number theorist by the name of Clifford Cocks was told of Ellis's ideas. While pondering them later that same day, he discovered in the space of half an hour what is now known as the RSA algorithm.

Early the following year, an old school friend of Cocks, Malcolm Williamson, joined GCHQ as a cryptographer. When Cocks described his scheme to Williamson, the latter studied it carefully for many hours, certain that he would find a flaw somewhere in the reasoning. He couldn't. Instead, he discovered another cryptographic protocol, now termed the Diffie-Hellman-Merkle key exchange.

These three men have not yet become as famous as their American counterparts. Whether they do or not seems to be a matter of indifference to them, which is something I find truly admirable. I wonder who was the unknown author of that Bell Telephone report.

Giving Nothing Away

Before we left the subject of cryptography, Dad talked a little about some other cryptographic protocols that he thought might interest us. These do not concern enciphering and deciphering, but issues where secrecy is also a requirement. First he told a story, which I'll tell you in a moment, about a young girl solving a certain problem in the company of older, more knowledgeable people—a story which, he said, should encourage anyone just to think. "Do not be intimidated," he told us. "Do not adopt negative attitudes like, 'How could I possibly solve this when I know so little about the topic?' Try thinking more along upbeat lines, 'This sounds interesting, it can't be all that difficult, I wonder how this might be done. . . .' " Dad suggested that sometimes we're better off not knowing whether a problem is (supposed to be) difficult or not, like the student who, arriving in a lecture hall shortly after a lecture had ended, took down the two problems remaining on the blackboard and later handed in their solutions, unaware that the problems were meant not as homework but as examples of problems (up to then) unsolved. Does this true story sound familiar? If you've seen *Good Will Hunting*, it should. But here, without further delay, is the story about the young girl.

Average Wage

Some years ago at a party, a group of mathematicians were asked, "How might you determine the average earnings of a group of people in a room (at a class reunion, perhaps) without any individual's divulging his or her salary?" While the gray hairs went into overdrive pondering this, the young daughter of one of them piped up, "I know how." Now don't read any further until you try for yourself.

Ready to match your solution against hers? "Let me whisper a big number to the first person, and then let him add his salary to this number and write the answer on a piece of paper and pass it to the next person. This second person should add his salary to this new number and write his result on a *different* piece of paper, which he should pass to the third person. He must keep hidden the piece of paper he was given. Continue doing this all the way around the room until you reach the last person. Make sure that each person writes the answer on a piece of paper different from the one handed to him or her. When it gets to the last person I'll take the result and subtract my original big number to get the total of all their earnings. Then I can get the average."

This clever idea is known as a *zero-knowledge protocol* because it reveals no knowledge about any individual's salary.

How to Share a Secret

We are often asked, "Can you keep a secret?" But how do you *share* a secret? And why might you want to do such a thing?

Alice, Bob and their friend Claire are working on a wonderful project and things are going really well. They are making great discoveries and are being as good as gold keeping detailed records of all their findings, though this is especially hard to do when they are just bursting to get on with experimenting to find out more exciting facts. All their important documents are kept, along with whatever else is vital to their work, safely locked in a cabinet out of harm's way. They are very fired up and would work night and day if it were physically possible. Up to now they have been very fortunate to be able to work as a complete team at times which have suited all three of them. Now, however, because of circumstances beyond their control, this will no longer be always possible. As it happens, some two of them, though not always the same two, could still meet. They agree for the sake of the project that at any one of these times the two people available should continue with the work and that the absent partner be briefed at the next meeting. However, they insist that no one of them should ever work alone, and though they are friends and trust one another, they agree to put themselves beyond temptation by locking the cabinet so that *any two* of them can open it but no one of them ever can. So how do they do this?

This is worth thinking about. Maybe I should tell you that they use more than one lock on the cabinet and distribute keys to these locks among themselves in a clever fashion. It is quite easy. When you come up with a method for doing this you will have implemented what is known as a *(2,3)-threshold scheme.* Read no further until you have tried.

Here is what they do. They lock the cabinet using *three* locks, to each of which there are only two keys. If the locks are labeled L_1, L_2 and L_3 then Alice keeps a key to $L_1$and L_2 (but not to L_3), Bob keeps a key to L_2 and L_3 (but not to L_1) while Claire keeps a key to $L_1$and L_3 (but not to L_2). All six keys are used, and each person has two keys. Since each of them is missing a key to one lock, no one can ever open all three locks on his or her own. But any two have, between them, a full set of keys to all three locks, and so can open the cabinet.

You might like to help Alice, Bob and Claire figure out what to do if they wish Denis to join their team and still maintain a threshold of two, meaning that any two of the team of four can work without the others being present. In other words, they need you to help them come up with a (2,4) scheme. Will they have to buy more locks, or can they get away with making more keys to the locks they already have? It is not so obvious how to implement this scheme with locks and keys. (See Appendix B, page 304, to verify your answer.)

What if we increase the number of people to 11, say, and the threshold to 6?

A (6,11) scheme: *Eleven scientists are working on a secret project. They wish to lock documents in a cabinet such that the cabinet can be opened if and only if six or more of the scientists are present. What is the smallest number of locks needed? What is the smallest number of keys to the locks each scientist must carry?*

Now things get a little bizarre. This delightful problem is cited at the beginning of a paper by Adi Shamir, the *S* in RSA, with the title "How to Share a Secret." It was in this short paper that Dad first learned about threshold schemes and their uses. He read it to us in sections, saying when he was finished, "Now that is how a paper should be written."

The surprise is that, without having to know how to go about it, or in fact knowing whether it is possible, it can be argued that *at least* 462 locks are needed with each scientist carrying *at least* 252 keys.

Can you imagine this, just for eleven scientists? Apart from the fact that they couldn't carry all these keys around on their persons, they would spend all day opening (and shutting) the cabinet, which would look like nothing on earth with 462 locks attached to it. So much for this mechanical solution!

Having beautifully illustrated the impracticality of implementing such a cumbersome scheme, the paper proceeds in a clear and direct way to promise an alternative clean electronic solution based on relatively simple mathematics. Rather than use locks and keys, the idea is to share some secret number *D,* such as a safe combination or cryptographic key, by breaking it up into several "pieces." These pieces (called shadows) are other numbers related to *D* which on their own reveal nothing about it. Each member of the group is given a piece. In order to recover *D,* a certain minimum number (the threshold) of people from the group must combine their pieces in a certain way. Any number of people below this threshold cannot find *D* using their pieces; it lies beyond reach.

The digital signing of checks is one application of such a threshold scheme. The idea is simple. Suppose you are the president of a company but cannot always be available to sign some of the important checks that must be issued from time to time. What do you do? Give one or all of the five vice presidents the power of signature? No, you are not that trusting. You arrange matters so that a check is considered signed only if it has a certain number or digital signature *D* printed on it electronically by a special device. You use a (3,5) scheme to create five shadows of *D* and have them printed on magnetic cards, and you give each of the five VPs a different card. When any three of these are read by the device it temporarily reconstructs *D* using the simple but clever mathematics behind the scheme, and signs the check digitally. Once this is done, the device destroys all traces of the number *D* and the numbers read from the cards. Two VPs cannot collude to defraud the company, for their two pieces alone are not enough to generate *D*. So this scheme affords a certain measure of safety. It is also very flexible. If you promote a new person to VP she, too, can be given a new shadow by simply extending to a (3,6) scheme. If you trust one VP, say your daughter, more than the others, or want to give her more (but still not total) authority over the signing of checks, then you might give her a second shadow.

At this point I have said just about everything I wished to say about cryptography, and, of course, a whole lot more than was absolutely necessary. Even if you have acquired only the barest understanding of the central notions and terminology of cryptography, you'll have no trouble following the rest of the story. From here on, it's mainly nonmathematical in nature: I'll share with you the ideas rather than the math behind my two cryptography projects—and the fun and excitement I had presenting and defending them at different exhibitions, as well as what I learned from these experiences. I shall tell you just a little about the extra math I had to learn when I began to investigate a new public key system in my second project, but no more than you need to appreciate how my proposed system hoped to compare favorably with the famous RSA system.

Part III
Exhibition Time

10 Young Scientist '98

As the time for Esat Young Scientist '98 drew near, Dad thought that I should have a few practice runs at explaining my project. It would help me to get my ideas clear, and was bound to improve my verbal presentation. The project contained many ideas that would need to be made simple and interesting. I would stand at the small blackboard in our kitchen and begin talking, and every so often Dad asked me a question as Mom listened. She would say afterwards at which points she felt I didn't explain something well enough or I went too fast. I knew this was a good idea because I had no public speaking experience of any kind, but I hated these sessions with a vengeance. They made me feel so self-conscious. I did three or four of these mock presentations and each time the hardest hurdle was to begin speaking. I suppose it was because it all seemed so fake, but these sessions were very helpful and made me feel better prepared.

What I much preferred was going for half-hour walks with Dad along the quiet country roads near our house. Here I could talk without being looked at or having to look at anyone. I could concentrate much better this way. Mostly I talked and Dad just listened. I wanted to get the sequence of ideas ordered correctly, to sense their general outline and to distinguish what was important from what was more technical detail. All this helped me to prepare a verbal presentation of about fifteen minutes in length. That is roughly the amount of time allotted to a contestant to describe his or her project to a judge. It would be terrible to run out of time without having got to the really interesting stuff. The more I explained the ideas, the more I could see what questions people might ask. So I thought a little more and understood more. Dad would also ask the odd question. "Better I ask you questions tougher than anyone else will." These walks were very pleasant and got me out into the fresh winter air doing something healthy.

The hardest part of the project was writing it up. It is such a long and tiring process because by now I was almost as much a perfectionist as my father. It was an exacting ordeal, and other things which should have been done much sooner were left literally to the last minute. The night before I was due in Dublin I just about finished putting together my poster presentation for the project and editing the slides I had prepared with the Microsoft presentation program PowerPoint. The following morning I was up at six o'clock to catch an early train. At the time it was both disconcerting and exciting that I had no idea of my project's standard or how it might compare with that of other entrants. I was afraid that some of the other projects might make mine seem trivial, which is something I wouldn't have liked. However, no matter how different my project might be, or whatever its standard, I was certainly going to learn a lot by attending the exhibition. Although I didn't know what to expect I did feel that I had done a lot of work, and I was pleased with the clear, simple written report which I eventually put together for presentation to the judges.

The Irish Young Scientist Exhibition, Royal Dublin Society, 1998

I had never even visited the exhibition before, so my first reaction was one of total "gobsmackification"—I was blown away! It was so colorful and noisy and there were so many displays. I felt like a child at a fairground. There were over six hundred projects in all, entered in three categories—Biological and Ecological Sciences, Behavioral Sciences, and Physics, Chemistry and Mathematics. My project was an individual entry (as opposed to a team or group entry) in the last of these categories. Of the three categories, junior (sixth class primary school, first and second years of secondary school), intermediate (third and transition year) and senior (fifth and sixth years), I was in the intermediate category. Before setting up my own display I had a look at some of the other exhibits. Those in the senior section seemed to me unfathomable, but over the following days I learned a little about them and became more relaxed. I never fooled myself into thinking that I had a chance of winning one of the two main prizes for individuals, but instead set my sights on a category award. The other contestants from my school did not seem to be as anxious as I was. Vincent Foley was an old pro by now—this was his third time. He tried to get me to relax while I was waiting for the judges, but I just couldn't. What would the judges be like? Would I have to do all the talking, or would they ask lots of questions? What kinds of question might they ask? Would they be interested at all?

There were three judging periods, one on the first afternoon (a Wednesday) and two more in the morning and afternoon of the next day. All the judges were professionals, either from academic backgrounds or industry, but none of the interviews with my three judges was the same. Each was new and scary, interesting and worthwhile. I felt I fared better in the second and third sessions than I did in the first. I was happy with the length of time that each judge spent with me and the interest each one showed in the subject of my project. The whole experience, from the preparation of my report, display and presentation, right down to meeting the other competitors and hearing about the other projects, is something I would never have missed. I'd encourage anyone who is interested in any of the sciences to participate in science contests, not for the prizes but for the excitement of working on something of your own choice and the atmosphere at the fairs. For me, the judges' interviews were the best experience—educational and motivating. It is rewarding to see all your hard work paying off when a judging session goes well or people compliment you on your display. Getting constructive criticism from one of my judges at Young Scientist '98 was the main reason I entered the competition again in '99. And from listening to other entrants' experiences with the judges and seeing their displays, I felt that I could do a little better the next time around by applying what I had learned. I just knew I had gained experience.

The contest was won by Raphael Hurley, another Cork entrant, whose project display was on a stand next to mine. He's a son of a friend of my father's and I grew quite friendly with him during the week. He had carried out a mathematical analysis of the game of Monopoly with the help of *Mathematica,* so our projects had something in common. I, too, was to be delighted with my result. During the main award ceremony on the third evening of the exhibition, Friday, I was so happy when my name was announced as the winner in the Individual Intermediate Mathematics, Physics and Chemistry category. The next day, special awards were given out, and my late nights before the exhibition spent cutting and sticking cardboard were shown to have paid off when I won the Display Award for the same section. But it was at a small ceremony just before the closing of the exhibition on Sunday that I was to receive my biggest prize. I was not even aware this prize existed, so I was taken completely by surprise when I was called up to accept the Intel Excellence Award. The recipient of this award would represent Ireland at the Intel International Science and Engineering Fair (ISEF) to be held in Fort Worth, Texas, later in the year. I could not believe it. This was a huge prize and I was totally shocked. While everyone else was packing up their exhibits just minutes after this ceremony, I had to let Dad take care of mine. I was walking around in a daze.

One consequence of my having won the Intel Award was an invitation to give a short talk on cryptography to my fellow classmates in the Enrichment Course in Mathematics. This is a special course given on Saturday mornings during term by some dedicated lecturers at University College Cork for school students with a particular interest in mathematics. Some of the older students in the class had been selected as possible members of the Irish Math Olympiad Team, so I knew I would be talking to some smart cookies. If this wasn't daunting enough, my knowing a few of them fairly well made the prospect of talking in front of them even worse. Having given a number of talks since that time, I know now that I always feel much happier speaking to a group of complete strangers, preferably ones I shall never meet again. It was the first time I had ever been a guest speaker and it taught me a lot—mainly that I hate addressing groups of people.

Days after Young Scientist '98, as I started to drop down a few clouds, I began to think of how I could improve my project for the ISEF. Foremost in my mind was the thought that I would have to do something extra to bring it up to what I imagined would be "international standard." I knew I was entitled to take it to Texas as it stood, but I felt it lacked a certain amount of originality, so I was keen to find some way of strengthening it. After one science exhibition I felt more confident and wanted to do my best. I discovered what kind of a challenge lay ahead of me when I read with awe descriptions of some of the projects that had been past winners. I knew I had a lot of work to do. While I was not required to expand my project for the competition, for me it was an absolute must. From that week on I was on the lookout for new ideas.

11 The Birth of a Project

Dad often says "the right thing is always the hard thing." When it comes to making tough decisions I've found this to be very true, and I remind myself of this advice whenever I'm tempted to take the easy way out. One hard decision, which I am now glad I made, was my choice of work experience during my transition year. This program is designed to give students an opportunity to work for a week or two in a company or an industry that they might like to join after finishing school or college. Many use it to get a foot in the door of a local business so that they can get a summer job there later in the year. I wanted, however, to make the most of this opportunity in a way which would benefit me much further into the future.

I decided to be adventurous and search the Internet to see if there were any cryptography companies in Ireland. I found the names of three, and after doing a little further digging I e-mailed my CV to the one I had picked as my first choice, Baltimore Technologies. Needless to say I took care to mention that I'd received my Intel Award for a project on cryptography. I remember the feeling of excitement as I wrote that letter. When a reply came with an invitation to spend a week with these information security specialists, I could not have been more pleased and frightened at the same time. But I was determined not to let an opportunity like this pass me by, and though I feared I might find myself way out of my depth in the company of top cryptographers, I made myself go. At the time I had had little exposure to what working in a computer company might be like, and I knew very little about what it is people do in most technical areas. It was not normal for students from our school to go beyond Cork in search of a work-experience placement, as transition year coordinators must visit each student to make sure that everything is going well. Since Baltimore is a Dublin-based company it looked as if I would be spared this visit.

Whenever I look back at my time at Baltimore I get a renewed sense of the excitement I felt while I was there. Working at my desk in the company of the many others who shared the spacious open-plan basement office with its lovely brick tiled ceiling, I fancied myself as belonging to a team engaged in important work. I reveled in the general hubbub, which I imagined was similar to the atmosphere of a very busy newspaper office, and I had a sense of being at the heart of things. There was no such thing as an hour for lunch. It was a very informal affair where I either ate in front of my computer or took the two minutes or so needed to go with some others to the shop nearby where we'd have rolls made up. I generally started at nine but found myself finishing later and later each day, gradually working my way from a five to five-thirty finish to one well after six. I stayed with my uncle Brian, a younger brother of Dad's, and his family in their house in the seaside town of Bray, about a half hour's ride on the Dart railway system. The daily commuting to and from was all new to me, and gave me a real feeling of working in the city.

Dr. William Whyte, their senior cryptographer, took me under his wing immediately and treated me as an equal. He gave me a draft of an unpublished paper by a colleague of his, Dr. Michael Purser, the founder of the company, who at the time was away on a fishing holiday. The paper explored a possible new public key cryptographic scheme based on mathematics that was unknown to me. My task was to try to write code to see whether I could implement its ideas and "get something working." Before I could get down to programming I had to familiarize myself with the mathematics. Dad got one or two frantic e-mails asking for the most basic and, whenever possible, concise explanations of terms. I know I could have sought William's help, and on occasion I did, but I didn't want to bother him too much. It's OK to bother dads but not busy bosses.

As *Mathematica* was the only programming language I knew,

William allowed me to use his ID to access the copy of the program on Trinity College's Web site. So I set to work. I read some of the paper, wrote a piece of code, read some more, wrote some more. Eventually, the program began to take shape and the mathematical ideas lost some of their strangeness. By the end of my stay and after much effort I had a working implementation of the system. I had also become intrigued by the mathematical ideas that had occupied my thoughts for the last few days.

William gave me truly challenging tasks and so spared me the fate of having to make tea and stuff envelopes. To top a very enjoyable but too short stay I got a magnificent reference from William, some of which I now quote as it describes the experience I gained better than I could.

> Given her background, when she arrived we wanted to give her a project that would challenge her and make proper use of her abilities. By its nature, this meant giving her work that was perhaps harder than she'd been expecting. I am pleased to say that she rose to the challenge magnificently. She worked independently on concepts that most students only encounter late in their undergraduate courses. She was perfectly happy to have the concepts explained to her and then work on her own to realize them. She took up our time only when it was necessary and asked focused, directed questions. It's been a pleasure to have her both as an office-mate and as a colleague, and we wish her all the best in the future.

I think this reference says as much about its author as it does about me. I was again to experience this same generosity of spirit a number of times from distinguished men and women. In Irish there is a saying, *Mol an óige agus tiocfaidh sí,* which roughly means "Praise youth and it will flourish."

When I returned home I thought, Wouldn't it be great if I could work on some of the ideas in Dr. Purser's paper? If I were to study and present a detailed discussion of them, it would really enhance my project and give it an original element. By "original" I don't mean my own ideas, but something not to be found in the current science journals. However, even if Dr. Purser would give me the go-ahead to use his ideas, it would still be better if I could add something new of my own. I remember talking to Dad about how the scheme described in his paper exploits a property known as noncommutative multiplication (meaning that $a \times b$ need not be the same as $b \times a$). Dad suggested that if I wanted a different setting in which to exploit this simple but very significant property I could use matrices because matrix multiplication is also noncommutative. I was all ears, though I knew nothing about matrices. I'll share with you in a moment a little of what I had to learn about them so that I can give you a glimpse of what it means for matrix multiplication to be noncommutative. I promise that from then on there will be no more explicit mathematics, only light explanations of mathematical ideas.

I became all fired up and decided that I would just have to pluck up the courage to phone Dr. Purser. Before the phone conversation I was shaking with worry about how he would react to my request. After all, he is a highly respected mathematician educated at Trinity College, Dublin, and Cambridge, and I was just an upstart secondary school student with the nerve to phone him at his home seeking permission to use his ideas.

I introduced myself nervously, apologizing for making so bold as to contact him. When I ran my request by him he answered, "Of course, fire ahead." I was so taken aback by this immediate reply (though it was the one I had wanted to hear) that I was lost for words. He was delighted that someone, particularly someone so young, would take an interest. I couldn't thank him enough. Hours later I was still thinking of the conversation I had just had and the fact that I could now work on something new and fresh. I guessed there would be an awful lot of work ahead of me, but I was very excited. I wrote a quick e-mail to William telling him of my phone call. An almost instant reply wished me the best of luck. I did not meet or get to know Dr. Purser until some time later. I now take the liberty of calling him Michael.

Two-by-Two Matrices

Matrices are an invention of Sir Arthur Cayley, a brilliant English mathematician of the nineteenth century. The array of numbers

$$\begin{pmatrix} 5 & 6 \\ 12 & 2 \end{pmatrix}$$

is an example of a 2×2 (read as "two by two") matrix. It is so called because it has two rows and two columns. It has four entries which, reading "row-wise," are 5, 6, 12 and 2.

As with numbers, it is often convenient to refer to a matrix by a single letter, for example

$$\mathbf{A} = \begin{pmatrix} 5 & 6 \\ 12 & 2 \end{pmatrix}$$

Another example of a 2×2 matrix is

$$\mathbf{B} = \begin{pmatrix} 4 & 5 \\ 3 & 9 \end{pmatrix}$$

Two matrices are added by adding their corresponding entries. For example,

$$\begin{pmatrix} 5 & 6 \\ 12 & 2 \end{pmatrix} + \begin{pmatrix} 4 & 5 \\ 3 & 9 \end{pmatrix} = \begin{pmatrix} 5 + 4 & 6 + 5 \\ 12 + 3 & 2 + 9 \end{pmatrix}$$

It is natural to label this matrix sum as $\mathbf{A} + \mathbf{B}$. Thus,

$$\mathbf{A} + \mathbf{B} = \begin{pmatrix} 9 & 11 \\ 15 & 11 \end{pmatrix}$$

Check that the matrix $\mathbf{B} + \mathbf{A}$ is the same as $\mathbf{A} + \mathbf{B}$. In general,

$$\mathbf{A} + \mathbf{B} = \mathbf{B} + \mathbf{A}$$

Matrix addition shares this "commutative property" with ordinary addition.

The rule for multiplying two 2 × 2 matrices is more complicated. Here is how the matrix **A** is multiplied by the matrix **B**:

$$\begin{pmatrix} 5 & 6 \\ 12 & 2 \end{pmatrix} \times \begin{pmatrix} 4 & 5 \\ 3 & 9 \end{pmatrix} = \begin{pmatrix} 5 \times 4 + 6 \times 3 & 5 \times 5 + 6 \times 9 \\ 12 \times 4 + 2 \times 3 & 12 \times 5 + 2 \times 9 \end{pmatrix}$$

Using the natural notation $\mathbf{A} \times \mathbf{B}$ to denote the product of the matrix **A** by the matrix **B,** this calculation shows that

$$\mathbf{A} \times \mathbf{B} = \begin{pmatrix} 38 & 79 \\ 54 & 78 \end{pmatrix}$$

This multiplication process may not make any sense to you at first, but there's a method to the madness. As a first step to explaining how the entries of the product $\mathbf{A} \times \mathbf{B}$ are calculated, imagine that it is divided into rows and columns:

	column 1	*column 2*
row 1	38	79
row 2	54	78

We might say that each is obtained by multiplying the appropriate row of **A** by the appropriate column of B. For example, the entry, 54 of $\mathbf{A} \times \mathbf{B}$, which is in the second row and the first column, is obtained by multiplying the corresponding row of **A** by the corresponding column of **B.** Thus the second row of **A,** which is (12 2), is multiplied by the first column of **B,** which is

$$\begin{pmatrix} 4 \\ 3 \end{pmatrix}$$

This row is multiplied by this column in the following manner:

$$\begin{aligned}(12 \quad 2) \times \begin{pmatrix}4\\3\end{pmatrix} &= 12 \times 4 + 2 \times 3\\ &= 48 + 6\\ &= 54\end{aligned}$$

What we have done here is multiply the corresponding entries of the row and column and added their results. Thus the first entry of the row, 12, multiplies the first entry of the column, 4, to give 48. The second entry of the row, 2, multiplies the second entry of the column, 3, to give 6. Their results, 48 and 6, are then added to give 54.

To calculate the entry 79, which appears in the first row and the second column of **A** × **B,** you multiply the first row of **A** by the second column of **B.** This is done exactly as described above. Here is the calculation:

$$\begin{aligned}(5 \quad 6) \times \begin{pmatrix}5\\9\end{pmatrix} &= 5 \times 5 + 6 \times 9\\ &= 25 + 54\\ &= 79\end{aligned}$$

See if you have got the hang of matrix multiplication by calculating for yourself the entry in the second row and the second column of **A** × **B**. It doesn't matter if you don't succeed—all that is important is that you can get the gist of what follows.

Now, when the matrix **B** is multiplied by the matrix **A** we have

$$\begin{pmatrix}4 & 5\\3 & 9\end{pmatrix} \times \begin{pmatrix}5 & 6\\12 & 2\end{pmatrix} = \begin{pmatrix}4 \times 5 + 5 \times 12 & 4 \times 6 + 5 \times 2\\3 \times 5 + 9 \times 12 & 3 \times 6 + 9 \times 2\end{pmatrix}$$

Again using the natural notation **B** × **A** to denote the multiplication of the matrix **B** by the matrix **A,** this calculation shows that

$$\mathbf{B} \times \mathbf{A} = \begin{pmatrix} 80 & 34 \\ 123 & 36 \end{pmatrix}$$

There is a surprise in store when we compare the two products $\mathbf{A} \times \mathbf{B}$ and $\mathbf{B} \times \mathbf{A}$ to see whether they are "equal." Two matrices are considered to be equal if, and only if, all their corresponding entries match (which seems a natural enough definition of equality for matrices). We find that

$$\mathbf{A} \times \mathbf{B} = \begin{pmatrix} 38 & 79 \\ 54 & 78 \end{pmatrix}$$

$$\mathbf{B} \times \mathbf{A} = \begin{pmatrix} 80 & 34 \\ 123 & 36 \end{pmatrix}$$

Compare the entries of $\mathbf{A} \times \mathbf{B}$ with the corresponding entries of $\mathbf{B} \times \mathbf{A}$—they do not match.

So, in contrast to the ordinary multiplication of numbers, for which it is always true that $a \times b = b \times a$, it is usual, as in this case, for matrices **A** and **B** that

$$\mathbf{A} \times \mathbf{B} \neq \mathbf{B} \times \mathbf{A}$$

(the symbol $\neq$ means "not equal to"). This is why matrix multiplication is said to be noncommutative.

Because matrix multiplication is noncommutative, things which are awkward in ordinary algebra become much more so in matrix algebra. For the ordinary numbers a and b,

$$(a + b)^3 = a^3 + 3a^2b + 3ab^2 + b^3$$

Here the more compact notation a^2b is being used instead of the longer $a^2 \times b$. The "expansion" of $(a + b)^3$ on the right-hand side of this equation is unpleasant enough, but it is positively compact by comparison with its matrix counterpart:

$$(\mathbf{A} + \mathbf{B})^3 = \mathbf{A}^3 + \mathbf{A}^2\mathbf{B} + \mathbf{ABA} + \mathbf{AB}^2 + \mathbf{BA}^2 + \mathbf{BAB} + \mathbf{B}^2\mathbf{A} + \mathbf{B}^3$$

The expansion of $(a + b)^3$ is not as extended as this because the commutative property, $a \times b = b \times a$, for ordinary numbers allows compound terms like $a \times b + b \times a$ to be compressed into single ones like $2ab$. Unless $\mathbf{A} \times \mathbf{B} = \mathbf{B} \times \mathbf{A}$, a similar compression is not valid for the matrix expression $\mathbf{A} \times \mathbf{B} + \mathbf{B} \times \mathbf{A}$.

It is precisely because matrix multiplication is noncommutative that attempts are made to exploit this feature to design cryptosystems that are fast and secure. Of course, arithmetic is performed modulo some positive number, which has the effect of complicating matters even further. For example, working modulo 13,

$$\mathbf{A} \times \mathbf{B} = \begin{pmatrix} 5 & 6 \\ 12 & 2 \end{pmatrix} \times \begin{pmatrix} 4 & 5 \\ 3 & 9 \end{pmatrix} = \begin{pmatrix} 12 & 1 \\ 2 & 0 \end{pmatrix}$$

while

$$\mathbf{B} \times \mathbf{A} = \begin{pmatrix} 4 & 5 \\ 3 & 9 \end{pmatrix} \times \begin{pmatrix} 5 & 6 \\ 12 & 2 \end{pmatrix} = \begin{pmatrix} 2 & 8 \\ 6 & 10 \end{pmatrix}$$

Besides adding a little more mystery, working modulo some positive integer has the virtue of keeping the magnitude of the numbers being handled less than the size of the modulus.

When using matrices, plaintext is broken up into blocks that are assigned numerical equivalents in the usual way. These numbers are then taken four at a time and one each placed in the four positions of a 2×2 matrix. For example, the plaintext message "buy shares now," which consists of exactly twelve characters, can be broken into the four blocks, "buy," "sha," "res," "now," each of size 3. An enciphering using trigraphs would then assign the plaintext message "buy shares now" the 2×2 matrix

$$\begin{pmatrix} 1220 & 12{,}350 \\ 11{,}614 & 9174 \end{pmatrix}$$

since these trigraphs have the numerical equivalents 1220, 12,350, 11,614 and 9174, respectively. This "plaintext matrix" is then transformed into a ciphertext matrix. The corresponding ciphertext is extracted by reversing the assignment procedures used in the encipherment. But could I come up with a system that would be faster than current ones but equally secure? Trying to find the answer to this question was to be the objective of my new project in cryptography and was one to which I hoped to have a positive answer in time for the Texas fair.

Heading to America

The Intel International Science and Engineering Fair in Texas was coming up in May. One abiding memory I have of the last few days of March 1998 is of spending a vast amount of time laboriously filling out forms. After having completed what seemed like 201 of them, I had to write an abstract of the project I was submitting to the competition. This was pure torture—it was particularly difficult as I had nothing of any substance done on the "new stuff." I was still reading and learning the basics. In the end I wrote a synopsis of my Young Scientist '98 project and just added the following ambition:

> I am currently doing some extra work which I hope to use to broaden my project. I am investigating the possibility of constructing an asymmetric algorithm which exploits the non-commutative property of matrix multiplication.

Because I had chosen a different setting in which to apply Michael's ideas, I would be on my own when it came to proving that the proposed new public key cryptography system was as secure as the one I was intending to compare it with—the celebrated RSA, which was the showpiece of my Young Scientist '98 project and which is accepted as a worldwide standard. If (and it was a big if) I could get the theoretical end right, I was hoping to show that the new system had certain speed advantages over the RSA. The way I was going to do this was simply to implement both systems (I had the RSA already programmed) on a computer and compare their "running times"—the times each would take to encipher and decipher an appreciable amount of text. I eventually ended up using twelve copies of the *Desiderata* ("Go placidly . . .") all strung together as my plaintext. This was going to be the easy but revealing part. Once I got the "new" system programmed correctly, it would simply be a matter of straightforward testing and recording, which I hoped would see it first past the post in the "speed" race.

Why did the proposed new system (assuming it was as secure) hold out the prospect of being faster than the RSA? Because it uses only multiplication, while the established system (known to be secure) involves an exponentiation step, which is much slower. This was my *hope*. The new system's security was to hinge on the noncommutativity of matrix multiplication and Michael's ingenious enciphering/deciphering scheme, and would be faster than the RSA because it avoided that slow step fundamental to the older system's security.

So here I was in April, about two weeks after my stint at Baltimore, counting a chicken as yet to be hatched. By this time I had learned enough about matrices to have the new system up and running. The good news was that the speed trials showed this algorithm to be a clear winner. This was something at least, but it would be nothing if the system couldn't be shown to be secure against all conceivable attacks. This was where the real work would have to start. That is where the difference lay between what I had done for my first project and what I needed to do for the proposed extended project. All the mathematics for my first project is well known. I simply learned what was relevant to the material I presented, and wrote code to demonstrate some chosen cryptosystems in action. Now things were completely different. They were on a grander scale, and Michael's ideas were new, as was the proposed setting, so there was nothing in textbooks relating directly to what I was hoping to undertake. The fact that I knew a certain amount of number theory and had assimilated some of the ideas of cryptography was obviously going to be a great help, but because I was dealing with matrices and no longer simply with numbers, I now needed to understand so much more mathematics. Normally one studies a topic in some depth before embarking on work which requires at least a basic knowledge of its content—but I didn't have the time! If I were to prepare in a normal fashion, I might do part of a university course over a number of months. My hoping to achieve everything I knew I would have to in a very short period was like insisting on writing a book in an unfamiliar language using only a dictionary, rather than learning the language first. Crazy, I know, but what the heck . . .

To allay my fears and put me a bit at ease, Dad showed me a passage from E. T. Bell's *Men of Mathematics* which had struck him forcefully when he first read it as a young man. It tells of the great German mathematician Karl Jacobi warning about dilly-dallying and encouraging young people to be brave and just get "stuck in" to whatever problem is at hand.

> Jacobi seems to have been the first regular mathematical instructor in a university to train students in research by lecturing on his own latest discoveries and letting the students see the creation of a new subject taking place before them. He believed in pitching young men into the icy water to learn to swim or drown by themselves. Many students put off attempting anything on their own account till they have mastered everything relating to their problem that has been done by others. The result is that but few ever acquire the knack of independent work. Jacobi combated this dilatory erudition. To drive home the point to a gifted but diffident young man who was always putting off doing anything until he had

> learned something more, Jacobi delivered himself of the following parable. "Your father would never have married, and you wouldn't be here now, if he had insisted on knowing all the girls in the world before marrying *one*."

Learning the basics about matrices wasn't as bad as I had imagined it might be in one of my many moments of panic. It was fun, and Dad told me about a few applications of matrices, which made them more real. The fact that I was going to be using them in a cryptographic system would be another application. I knew Lester Hill had used them in the 1930s in a classical cryptosystem. It was at this stage that I was able to program the new system and check the speed prediction experimentally. So the practical end was OK. I could try all sorts of variations and fine tunings whenever I just wanted to play.

But was the system secure? The only way to know is to envisage all sorts of attacks and then show (hopefully) that none can succeed. Even at this point, it may be that some subtle or not so subtle attack has been overlooked. Of course, you won't know this when you report on what you investigated and the conclusions that you reached. Another person might see a gaping hole in the system, but you hope that this won't happen, and if it does, you try to deal with it. So, in broad outline, the plan was to attack and defend.

In order to answer the many questions which such investigations threw up, I had to ferret out as much relevant material as possible, which required "going to the journals"; very specific subject matter is not to be found in general textbooks. This was a new experience and gave me a much better appreciation of the importance of good libraries. I remembered Dad saying that Gauss had chosen to attend Göttingen University rather than some other one because its library was better. I thought this a little odd at the time but I now understand it, even if it wouldn't be *my* overriding consideration in choosing a university. Initially with Dad's guidance, I spent hours in the libraries of the Cork Institute of Technology, where, as I've said, he works, and University College Cork, searching through the many mathematical journals and magazines for papers which might contain something that might shed some light. The papers which were to prove the most help to me came not from the "very heavy" journals but from the lighter ones, whose articles contain plenty of mathematics but are written in an easy style aimed at keeping the reader's attention.

Then, of course, I had to read the papers very carefully. In the beginning Dad went through one or two of them with me, pointing out what I should be able to read, and what I couldn't be expected to understand, given my current state of knowledge. He knew I would get a terrible shock, so he warned me not to despair and advised that I read a paper through once quickly (even if most of it was double Dutch) to get a vague idea of whether it might be relevant. Even with this encouragement it was still a chilling experience at first because I kept coming across terms I didn't recognize and results which were being taken for granted, but which I knew absolutely nothing about. This was a time of "deep realization." If I ever had any doubts about how much work I would have to do, I certainly had none now.

It was easy to deal with new terms—look them up in a standard textbook or just ask Dad; but the results that were being reported were an entirely different matter. "Dad, what is the Chinese Remainder Theorem?" might be answered by an explanation with a proof or one without, or it might simply be answered by, "You can look it up in such-and-such a book." One way or the other it normally meant at least a day's work just trying to figure it out. With a particularly interesting theorem, like the one I just mentioned, I'd write a program to generate examples. I find this a great way (when possible) of getting to know a result. Whenever I programmed something I'd end up playing around for hours rather than getting back to plodding my way through the paper. Eventually back to the paper I would go, to decipher (!) a few more lines.

In time it became a little easier to recognize almost at a glance when a paper might be useful and, much later, even easier to read a paper right through while all the time on the lookout for material which would be bang on for what I was doing. Those papers that held a promise of being helpful I read again and again.

All of this was an unusual experience for me, but I had a great feeling of excitement. I think it was because I was working on something that no one had worked on before. I worked constantly for whole days on end, and it was exhilarating. There were times when I never wanted to stop. Most of the time nothing was coming together and I was running into dead end after dead end. I kept different parts of the developing ideas on the blackboard in the kitchen, in particular those parts where I was stuck, so that they were there in front of me all the time, waiting to be sorted out. On one occasion very early on I had three laptops on the kitchen table all running different tasks in *Mathematica*. This was about the only time, I can say with certainty, that my preoccupations caught the boys' attention. Otherwise, they appeared as if they couldn't care less and were more than happy to leave Dad and me to our own private world.

Then there was the time when I came bounding down the stairs to tell Dad of a simple step which meant that everything I had done before would fit together to finish a large part of a proof. It was a real joy when inspiration hit, but I couldn't waste time with self-congratulatory backslapping. It was back to work on a full tank of enthusiasm.

As my understanding of mathematical ideas deepened, so, too, did my approach. I wanted to learn more and more and, thanks to Dad, I wanted to learn it properly. I knew that it is reasonable to quote results and not be expected (probably) to be able to know how to prove them. Knowing them is often enough, but I started wanting to know how everything is proved. I wasn't happy to plow ahead and search for answers to other questions and assume that what I have read or been told is true (which it normally is) without getting sidetracked. When I did read a proof where the math was far above my level of understanding I became frustrated and often wanted to learn the relevant math, even though I knew it wasn't necessary or directly related to what I was doing. I came to hate being told, "It's probably best understood by working through the proof, but it's too difficult." Everything Dad had taught me he taught me—either at home or in his lectures—from the bottom up, beginning with "the fundamentals, Sarah, the fundamentals!"

Just to give a sample of the kind of things I was working on from day to day once things got under way, here are some extracts from my diary for that period, which I have come to regard as "those extraordinary months." (Don't worry about knowing exactly what I mean by each entry.)

> Started slowly—experimenting with matrices over finite fields and finite commutative rings—Made routine for generating random matrices—Learnt matrix algorithm "Hill cipher"—Made own routine for finding inverses of matrices—Put together first enciphering and deciphering routine, only nine characters at a time though—Next step, get routine to work on lists of matrices—Wrote general routines after solving "join problem"—Found **Flatten**[ ] command at 23:43 tonight (has solved loads of problems)—Investigated orders of groups—Surfed the net for papers and notes on various things including discrete logarithm problem—Ran first "timing" using *Desiderata* × 12—Investigated possible signature schemes, a big learning experience—Learned more about rings, groups etc.—Quadratic residues—Got two laptops from CIT which could talk to each other. (For display purposes)—M. Vandyck taught me about eigenvalues, eigenvectors etc.—M. Vandyck gave me an overview of spectral decomposition theory for extracting the roots of matrices—Talked to Dr. Pat Fitzpatrick, told him all the details, very helpful—Went to UCC and CIT libraries and looked at math journals and number theory and linear algebra books—Found problem in routines (very silly mistake)—Programmed Hill cipher for interest—Solved lots of the problems in *Mathematica* stuff, put more restriction on different parts of the rou-

tines to stop any unforeseen problems with commutativity etc.—Got a fright in morning thinking that $\delta.\epsilon = \alpha.\delta$ would be solvable by linear equations but thankfully too many possible answers—A breakthrough.—Learnt Gaussian elimination from Dad—Got *Mathematica* to find all possible solutions for a few small examples—One-to-one mappings—Programmed row reduce.—Kept investigating how to find χ and number of possible χ's to fit certain equations—Went back to basics, testing myself, investigating groups of small order—Made a conjecture that the no. of invertible matrices in $\mathbf{GL}(2,\mathbf{Zn})$ 5 no. of invertible matrices in

$$\mathbf{GL}(2,\mathbf{Z}_p) \times \mathbf{GL}(2,\mathbf{Z}_q) = n\phi(n)^2(p + 1)(q + 1)$$

—how to prove with arguments like one for order of $\mathbf{GL}(2,\mathbf{Z}_p)$? Investigating $\mathbf{GL}(2,\mathbf{Z}_6)$ separating matrices of different determinants into groups—Investigating structure of subsets—Primitive elements—Listing "pre images" of quadratic residues—Wrote matrix routine to reduce modulo n at each step and tested this against the method I had been using for speed advantages—Reverted to old way after many tests—Testing against RSA—Plotted a few graphs of comparison times obtained empirically—Tidied up some files—Modified RSA routines to make p and $q \equiv$ 3 mod 4—Big change, changed routine to give a high-order element by making p and q both products of two times another prime plus 1—Everything coming together—the end is in sight!

Speaking math, writing math, thinking math nearly all the time was just so weird. I had a constant feeling of "I can't believe I'm working on this—I can't believe I'm working on math like this!" I was enthralled by the work and found it fascinating, so much so that for weeks I completely forgot normal pursuits such as seeing friends socially or going sports training. I came to understand

Dad's obsession and could finally comprehend how a book—other than a school textbook—could be written on mathematics. When I was younger I had often expressed amazement when Dad took down one of his many books and began to read it. "How can you read a book about math—what could you possibly find so interesting?" I had assumed they were the same as our schoolbooks, containing a few worked examples of a technique followed by hundreds of sample questions. The English in these schoolbooks is usually limited to words like evaluate, simplify, verify, prove, show . . . so maybe you can see why I was at a loss to understand how anyone would wish to read such stuff.

One thing I used to find a little annoying about Dad was his desire to know more about mathematicians. When I'd speak about interesting work I had just read by some mathematician or cryptographer, he would almost always ask, "Do you have a picture of him/her?" He says knowing a little about the people behind the math can often make it more interesting. Students especially love to be told an anecdote which lightens a lecture or helps them remember a particular topic. Although I don't wish to give the impression that I never found a little history of mathematics or mathematicians of some interest, I did think Dad's curiosity a little excessive. However, having painstakingly gone through mathematical papers written by men and women who are alive as I write, and having read many popular articles about mathematics and cryptography, I now have a much keener interest in the human side. Recently I saw for the first time photographs of five men of whom I have heard so much and even talked about a little myself—Martin Hellman, Whitfield Diffie, Ron Rivest, Adi Shamir and Leonard Adleman. I must say that I was a little amazed to see these "giants" in normal bodies. Now I hate to hear about the work of a mathematician without also being told a little about the person, or seeing a picture of him or her. I am going to have to do something radical to stop myself from turning into Dad!

In the end, the papers that helped me show that the attacks I envisaged were not computationally feasible turned out to be almost the very first ones I had read. But it took a long time before

I could see clearly exactly what I had to do. I did an awful lot of jumping around and I indulged in many programming diversions. There were times when I could hardly think straight and often asked questions of myself that afterwards I could see were plainly stupid. But it all came together, culminating one evening in a "That's it!" which had me leaping with joy. Something had finally dawned on me, though it had been staring me in the face for a long time. I called a halt. I knew there was a lot more I could do, but I'd had enough, in both senses: though I was mentally drained I now had a genuinely good extension of my '98 project. Texas, here I come!

I had never been to the USA before, so I was really looking forward to my visit to Fort Worth accompanied by my science teacher, Mr. Foley. Intel paid for everything, which was absolutely great. We stayed in a hotel where I had a king-sized room to myself for the week. My project was deeper, thanks to Michael Purser, and I just knew I was going to enjoy myself. The contestants ranged in age from fourteen to twenty-one. I was in the Mathematics category, and my stand was between those of two American boys. The fair was completely different in character from the Irish exhibition. Although it was definitely an international event, most of the contestants were American. In general, I found the participants to be very confident when speaking about their projects. Many of them, however, seemed to me to be overly so, trying too hard to sell what they had done. Others had an air of knowing more than they did, and didn't fare well when they were asked more probing questions. Some projects were obviously brilliant and did not need to be sold—they spoke for themselves. The displays were all very professionally done, and about twice the size of the ones at the Irish Young Scientist Exhibition. There were at least a thousand contestants and, it seemed to me, about as many judges.

The biggest difference was in the judging process. There were nearly as many judges as contestants, so the system was a complicated one. Each contestant received an appointment card with each fifteen-minute section of the judging day marked either "open" or with a scheduled judge's name. Special award and additional regular judges could be expected during open periods. A bell rang at the end of each period, so there weren't going to be any judges talking to you for over half an hour, as they had been able to in Dublin. The morning before the first judging sessions, Frank Turpin, the "man with the money" from Intel, took me shopping to take my mind off things. It worked, and we arrived back at the convention center ten minutes before the judges started making their rounds at 12:15. I was not due a scheduled judge until 1:30, but I was judged right from 12:15 until 5:30—no lunch break, no nothing! There was only one day of judging, and it was hectic. The questions asked were wide-ranging and varied. I was even asked about the difficulty of factoring large *primes*!* I'll never forget that! It was the very slip of the tongue that I myself had made in one of my practice sessions. Dad had done a flip, saying that I'd be ruined if I ever made this mistake in public. At one point another judge came to see me with a notepad on which he had a whole list of questions written out. The judges had obviously collaborated before "sending him in." The questions were tough mathematical ones designed to make sure I was not bluffing. He got straight down to his task. I wrote out proofs for this and that, and explained certain concepts. When he went away smiling, I knew I could smile too. From the number of judges I had received I knew something was in the offing.

At the awards ceremony I was given a third-place Karl Menger Memorial Award from the American Mathematical Society and a fourth-place Grand Award in Mathematics, and I was very happy—but there was more to come. I remember laughing as the girl next to me jumped up into the air and ran towards the stage in a burst of excitement when her name was called out for a major prize. Mr. Foley had also noticed her leap and we both laughed for

*In case you'd forgotten, we don't talk about factoring a prime because a prime has only itself and 1 as its factors. It is OK to ask, "What are the factors of the number 889,363?" but it is not very sensible to ask, "What are the factors of the prime 889,363?"

her and at her show of delight. However, I certainly would have beaten her had there been a competition for who could jump the highest off a chair from a sitting position, because my name was called out next. I had felt so tense through the whole ceremony wondering if I would get any prizes, and since I had received two already I certainly wasn't expecting any more in the final few minutes. To my complete surprise, at a time when I thought I should be calming down, I instead found myself practically flinging my jacket at an innocent bystander who had been minding his own business near the steps to the stage. I was beaming as I received the $2000 Intel Fellows Achievement Award, and smiled as I thought of the two leaps that my fellow prizewinner and I had made just moments before. The evening and my trip to Texas were complete after that. I felt absolutely happy and content.

The minute I got back to my hotel room I rang Mom and Dad with all the news, delivering it, as ever, at a hundred miles an hour. I told Dad that the Americans were making a fuss over the "new" algorithm, and I asked him to phone Dr. Purser to tell him of the great reception I had received in Texas. I also wanted to give Michael credit by naming the algorithm the Cayley-Purser algorithm.

Up to this point, I had simply called the algorithm the Cayley algorithm because of its use of matrices. I had no idea that it would create a stir—for me it was just a nice mathematical project of which I felt reasonably proud and comfortable because it had more than merely textbook ideas. When people noticed it, I thought it only fair that I give credit to Michael Purser. Of course, the addition of the name of someone still alive gave the project another human dimension.

When Dad told Michael this he said, "At my age I don't need recognition or awards." He was very kind, and remarked that he was flattered at having his name linked to that of Cayley, saying this would be the second time he had been mentioned in "such distinguished company." An author friend of his, having expressed a certain view in his book, had added, "though Einstein and Purser will disagree with me." From that point on I always referred to the algorithm as the Cayley-Purser algorithm, or the CP algorithm for short—C for Cayley, the inventor of matrices, and P for Purser, the man who gave so generously of his ideas.

Preparing for Young Scientist '99

Shortly after arriving back in Ireland I visited Baltimore Technologies with Dad. We were met by William, who introduced me to Michael, the man whose ideas I had worked on incessantly for weeks but never met until now. I had brought two copies of my project report with me and gave one to each. Michael, who had a copy of his book *Introduction to Error-Correcting Codes* in his hand, on hearing that Dad lectured on this topic, immediately signed it and gave it to him. We all went for tea and cakes (which I was told William had gone to great lengths to procure), and I met again all those people I had got to know during my week's stay.

William and Michael were able to see the work I had done. I was extremely flattered by the compliments they paid me by e-mail when they had read the report. But what delighted me, and was much more important to me, is that they took a keen interest. Michael in particular wrote many times with various suggestions.

I spent almost all of the summer on my grandparents' farm outside Killarney helping my aunt with the horses. In September, I reluctantly returned to school for the fifth year to begin the two-year cycle for the Leaving Certificate examination. I had done no further work on my project other than think about a number of plans I had for tying up loose ends which I hadn't been able to do before going to America. At Young Scientist '98, I had been encouraged to expand and improve upon my project for next year's competition. Clearly I could now do this with the material I had worked on for the Texas fair. However, rather than just expand it I decided to shed almost all of the old project, except for the portion that described the RSA. This, along with the new material, would be the basis of a fresh and deeper project that would focus solely on comparing and contrasting the RSA algorithm with what I was now calling the CP algorithm. The title of this project was to be "Cryptography—A New Algorithm Versus the RSA."

I was very pleased that I had kept records of all my work because they would help get me back to the frame of mind in which I could begin to tackle those questions which were still unanswered. I had two other ambitions: one to write crisper code (I knew my early stuff was primitive in many respects); the other, if possible, to include a "complexity analysis," which I now explain.

In my e-mail correspondence with Michael and William there was a letter from Michael discussing the possibility of explaining theoretically a figure I had obtained experimentally: the ratio of the times the CP algorithm and the RSA required for enciphering and deciphering. I found it intriguing that by thinking carefully about what an algorithm did while it was carrying out its tasks, and by making educated guesses as to how long each of the steps takes, one could hope to explain figures which were the results of practical tests. I imagined it to be almost like theoretical physics. Wouldn't it be fantastic if I could give reasons why the figures come out the way they do? So I was going to have a stab at this complexity analysis, though I had no illusions that it would be easy. It wasn't.

Circumstances helped to make writing better code almost painless. Dad was teaching the Mathematical Excursions class again and this year he was using *Mathematica* as a teaching aid. He had many examples prepared which showed all sorts of commands being used in very powerful ways. Seeing some of these commands used in ways slightly different from those I had used made me realize where I could make some of my procedures more efficient. I devoted much time to dissecting superb code Dad had taken from programming books, and admiring how cleverly it achieved its goals. I also began to look at code written by others, something I had done before but had found a little depressing because everything I read was simply too sophisticated. Understanding neater and cleverer ways of doing things is a great thrill, even if it does make you feel you are thick. I remember really admiring some code written by Stephen Kauffman of the ETH (the Eidgenössische Technische Hochschule, the Swiss Federal Polytechnic) in Zurich, which Dad had come across while at a *Mathematica* conference in Chicago in June 1998. It was written to implement the RSA system. Seeing how he tackled the enciphering and deciphering in a much more structured way showed me the wonderful flexibility of the *Mathematica* language. However, I often prefer to use my own code, even if it's not state of the art, because I believe it corresponds to my way of thinking. At least then I have some chance of following what I wrote weeks ago when I reread it. Of course, I do use and adapt ideas that make crude code less so.

I did a lot of work that I never reported because I didn't complete it. I failed to get the complexity analysis to a point where I could draw conclusions by which I could confidently stand. I wanted to include all that I had done to show how hard I had worked, but in the end I kept the report, like the code, as crisp as I could manage. Despite not winning all the battles I had an original project, and with my thoughts clear in my head I felt ready and eager for Young Scientist '99.

12 Young Scientist '99

To give you a sense of how a contest's excitement and tension gradually mounts as the days go from judging sessions to the awarding of prizes, here are my experiences and reflections straight from my diary entries:

> **Wed. 6th January 1999.** Mr. Foley and Vincent collected me from home close to 7:45 a.m. As usual, they were calm, but this was not how I was feeling. Normally I like getting excited about things, but on the journey to Dublin I was more tired than excited. I knew that the first judging session would be in the afternoon, so I was asking myself questions in my head and trying to anticipate what might really be asked. Got to the Royal Dublin Society (RDS) around noon. The hustle and bustle started the minute I arrived. Registered. Once I walked in the main door of the hall the atmosphere was overwhelming—a wave of noise and bright lights washed over me and I was filled with anticipation. It was very fun and professional at the same time. I felt as if I should be really cool and calm—after all, I was in the senior section this year and had been here before. However, I was even more excited than last time, though I wasn't as nervous, as I knew what was supposed to happen. I set up my project and said hello to the people around me. I found out that Hugh Hurley's project was on the other side of my stand (Hugh's brother Raphael was last year's winner) and Vincent's was just two up from me, so I settled down. I set it up slowly and, again this year, it stood

out as quite different. Some of the displays were really well done.

My first judge arrived earlier than expected, and I was disappointed that I wasn't more ready or tuned in. I had been talking to some of the other people at the stands around me and didn't have any time to myself to concentrate. I like to feel prepared, even if only the slightest bit. The judge was from Siemens*—Dr. Supple, a very nice man. I got the impression that he was not too interested in cryptography, but luckily it is a topic that can easily spark some interest in anyone if you try hard enough to explain what it is all about. I had a short introductory spiel prepared, which I gave him. After being in America I had a better idea of how to mix their exuberant methods of explaining their projects with my idea of a good solid explanation—cover the most relevant points and expand on those which seem to interest the judges. Dr. Supple stayed for about forty minutes, which was quite a long time. Even so, I don't feel I impressed him that much and I am disappointed that I did not get the chance to get into the nitty-gritty of the math which, after all, is the core of my project. It was a good session overall but unfortunately not too probing. I want to show the judges that I can answer (or so I hope) whatever they throw at me, and I hope that their questions will be mathematical in nature. Presumably the judges will understand that I do not have a knowledge of all the cryptographic ideas that have ever seen the light of day. I am always willing to learn and see no problem in the fact that I do not know as much as I might were I to read more widely. On the math side of things I feel OK, though I know I could be asked questions which I cannot answer. It is hard to know to what depth to go.

*In their own words, "One of the largest electrical engineering and electronics companies in the world."

At Young Scientist '98 I noticed that when people who knew something about cryptography came to my stand it was not with the aim of showing me up. Instead, they seemed anxious to show that they knew something about the topic. When they'd ask me what I thought of this latest development or that new idea, instead of attempting to answer (which sometimes I wasn't able to) I simply asked the person to elaborate. Suddenly, they would relax and become enthusiastic. This was a very interesting experience and a valuable lesson, though when a judge did something similar at last year's competition I found it a little unsettling. On one occasion a judge had proceeded to ask me questions about Enigma.* When I explained that it was not something I had studied in depth, he started telling me about it as he settled himself comfortably into a chair. Let's just say that the middle of a judging session was not when I wanted to hear all this. He also asked me, "Do you really think that Julius Caesar invented or even used the Caesar cipher?" Now, that was one that even Dad, who I thought had asked me every question under the sun, hadn't thought of. I was glad when the judge eventually let me get back to talking about my project.

Thursday 7th. During breakfast, all I could think of was, "I am going to make an impression today." *Wenn schon, denn schon.* If you do something, do it right. I wouldn't disregard the modesty that my father and mother had instilled in me, but I would do my best to show the judges what I knew and what my project was all about. I really psyched myself up. It felt weird. Yesterday I had adopted a laid-back attitude, taking the judging as it came and not doing anything major to make it go where I wanted.

*The cipher machine used by the German government and military just before and during World War II.

When we got to the RDS just after nine, a wave of tiredness swept over me because I didn't get to sleep last night until about four in the morning. It doesn't help that I have a very bad cold. I cleared away the copies of the previous day's edition of the *Irish Times* which had accumulated on the chair beside my stand and was about to sit down to clear my mind—yes, "clear my mind," the very phrase I hate Dad using—when I was met by a palm ready to be shaken and a very enthusiastic face indeed. It was my first judge of the day. I was suddenly awake. All my intentions since breakfast were fresh again at the top of my mind. The judge was Dr. Eoin O'Neill, and I was delighted to see him. He had judged me last year and had given me the most satisfying grilling on every aspect of that project. He had even spoken to me again after the judging sessions were over, to encourage me to re-enter the following year, and he had given me a few tips. So this year I was going to show him that I had listened to his advice. I wanted to knock his socks off! I know it was the mathematics that impressed him last year and the fact that I know the basic concepts of public key cryptography. This meant that I could get right down to the main business of comparing and contrasting the new algorithm with the RSA. I could tell by his face that he was looking forward to hearing what I had to say, and that he would listen to every single word.

But then I realized that there was another palm waiting to be shaken and it belonged to a man whose face was unfamiliar. He looked as though he was aware that my judge was already here, yet he made no move to leave Dr. O'Neill and me alone. Dr. O'Neill introduced him simply as someone he had brought along with him as an external mathematician. So he was here to stay, and he was going to listen to my every word. Boy, was the heat on now! It felt really scary, yet a massive challenge to which I was dying to rise. I felt I had to live up to Dr. O'Neill's obvious expectations, so I just started. Damn, it felt good! It was an excellent session. For over forty minutes I spoke as intensely as they listened. I wrote out each and every step of my beloved proofs on an A4 pad. I knew it would impress them more than if I simply slapped my report down on the desk and took them through it. I used Dad's horrid hand movements to explain certain concepts like homomorphisms and could see that they were watching every move I made. They listened intently as I wrote and explained the ideas and each step of my argument. At one point I just couldn't remember what I wanted to say next and told them I would get my report to show them some tables. Tables—that's it! I said, "Just before I do, I'll show you here on paper where I get what is in those tables." I could tell that this was one of the moments that impressed them most. I was delighted.

On one occasion I looked out of our little huddle and it felt really strange—our conversation was so very intense that just to look around was like coming up for air. I got a glimpse of Mr. Foley, who was looking dead impressed and dumbfounded at the same time. I hate having people I know listen to me explain to others something I have already talked about with them. For a split second I felt very self-conscious, but once I leaned down on the desk again I was sucked back into our world. This must have looked very odd to anyone looking on: three of us all bent over, talking about math proofs. They asked great questions (nothing about the difficulties of factoring large primes!). I delivered answers confidently and led them into other areas by expanding on certain points. I even talked about things I hadn't been able to figure out, such as the complexity analysis. Before they left, Dr. O'Neill asked me the simplest question of all, and I could see that he was wondering whether or not I would be able to answer it. The answer was the fast exponentiation algorithm, and I must have smiled before I replied because I knew it was the perfect end to the perfect session. I had been able to defend my project at all levels. The last question was a check to see if I knew the fundamentals. They smiled at each other on my final answer, which I'll never forget. We shook hands again, and I hoped no one would get in my way as I ran to the phone booth, red-faced from exhaustion and excitement, to let off steam in a phone call to Mom and Dad that cost me an absolute fortune.

In the afternoon my third and final scheduled judge, Dr. G. Frank Imbush, arrived around 3:30. His name rang a bell—I remembered Vincent talking about him last year, saying that he was one of the good judges, tough but fair. I didn't go into the mathematical side of the project on Dr. O'Neill's advice of this morning, which was just as well as I have never been so mentally shattered. I still wanted to make an impression, but I was just so tired. The session went well but afterwards I felt that, of them all, this morning's was the only really good one. I feel badly about the other judging sessions because they were of a different standard. And yet I am not too worried, though my tired performance of this afternoon could, I think, have jeopardized my chances of a prize. But the session with the two judges was just so damn good—I still feel tingles of excitement thinking of it. It made me realize what a good intellectual conversation can be like—this morning's was one of the best ever and it was about math. I am ready to go home, happy that I have done my best.

This evening I went to McDonald's with Grandma and Granddad, and then to see the film *The Mighty* in Stillorgan. Afterwards I met friends, but all I could think about was that interview with Dr. O'Neill, and that he had brought along an expert so that together they could ask me better questions. He is a good judge. What an extreme compliment that gesture paid me. I still don't know if I have lived up to his expectations. At the moment I feel that no one else could possibly understand the joy of such an interview or appreciate just how good a feeling I have about it. It is a lonely feeling but I don't mind. As long as I am happy within myself, that is more than a just reward.

I know that people cannot be expected to understand how I feel about the project, and it is probably for this reason that I do not like to be asked simply, "What is your project about?" It is not that I don't like to talk about it, but it cannot be explained fully in a few sentences. If I think someone isn't going to feel the same way about it as I do and just wants a superficial explanation, then I feel awful about trivializing it. That is not to say, however, that I do not respect or understand that not everyone gets a thrill from cryptography ideas. I like it that way—it makes it more special.

I forgot to ring Mom and Dad about that third judging session because I was so obsessed with the second one, which I still can't get over. I'll always remember that parting smile!

Friday 8th. Today I got more judges. If you get judges on the third day this means you are in the running, as they say, for a prize. This morning I had no such expectations—last year, when I won my category, I did not have any judges on the last day. So I began to feel quietly confident, thinking that, after all, this year's project is far better than last year's. The only point on which I still felt pessimistic was that I did not show what I feel to be an even level of ability when describing the project, so I thought a lot would rest on how respected Dr. O'Neill's opinion was amongst the panel of judges. When Dr. Supple came back with two more judges from other categories I was pretty pleased. He invited me to explain my project to them, which I was delighted to do. I had to express its merits without going into detail about what exactly they are. I explained things in simple terms—I had some straightforward props, such as the Caesar wheel, and I think they liked that in particular. I believe that nobody really minds being told something in a simple way. It is not an insult to anyone's intelligence and ensures that no one misses the point.

Later, Dr. Supple came back with three more judges, and I knew that things were looking good. I took all this attention in my stride, though last year I would have died for it when I was sitting next to the boy who won. The first of this set of judges introduced himself as being from the group section—though initially I thought he said group theory! Luckily I found out otherwise before I had him completely confused by speaking to him exclusively about my beloved proofs while appearing to ignore the others. It was a good session, but I was disheartened for a moment when one of the judges started looking around while I was speaking. In America they requested that judges appear to be interested in a student's project even if they think it to be trivial or uninteresting. They point out that the judging periods should build up the student's confidence in speaking on a topic in scientific or layman's terms. If one of my main judges had shown no interest I would have been deeply hurt. When I saw this particular judge look around I became distracted, and he upset my flow of thought and lessened my level of self-confidence. However, he seemed to tune back in again, and after a few nice questions on nothing too probing, the group left. I had looked over at one point during this interview and saw that my schoolmate Vincent too had judges. That was great! His father was leaning by the wall of the exhibition hall and was beaming with pride as he watched his son explain his project to his audience, which was obviously captivated.

Next thing I knew, judges from Intel came over. Then a judge from the Patent Office, followed by Dr. Scott and, a little later, by Rev. Thomas Burke (who advised me to read Simon Singh's book *Fermat's Last Theorem*). A few more judges called by, but I can't remember who they were. I know I had two judges from the Social and Behavioral category. While all this was happening Vincent also had more judges.

At 3:00 p.m. at least thirty of us were called backstage to wait for the closing ceremony, at which three individual and two group prizes would be awarded. My friend Raphael Hurley (the Young Scientist '98 overall winner) was also backstage, and he pointed out to me that besides myself there were only two other contestants with individual entries. All the others were group entrants, so according to Raphael most of them were "spongers," which was his way of saying that they were there to fill out the set of possible prize-winners. I didn't want to let myself believe that I was going to be one of the individual winners. When we were all called into a corridor right behind the stage, the televised ceremony got under way. Soon they would begin announcing the prize-winners. It was then that Vincent was called up to the side of the stage and told that he was about to receive the Intel Award. A few moments later I was also called up, and Vincent told me his wonderful news. For one awful moment I thought I was only there as last year's Intel winner to present him with his prize. I was delighted for Vincent, but felt crushed inside. When Karina Howley (an Intel lady I know from last year) came over to the two of us and congratulated Vincent and then me, I went, "Hey, Karina, I don't know anything yet!" Two other guys from Intel, David and Ian, were beside the stage chatting away to me when an Esat girl came and took me backstage again to where all the "groupies" were now settled. Vincent returned from his stint in front of the cheering audience. He was dead proud! Part of his prize was a trip to ISEF '99 in Philadelphia to be accompanied by his Dad as his teacher.

Karina's comment had set my mind spinning with excitement. Then Hugh Hurley and his teammate Conor were called for the runner-up group prize. Next, two girls from Belfast were called to receive the overall group prize, followed by David Folan from Donegal as the individual runner-up. At this point I was the only individual entrant left backstage. I could hear the Taoiseach speaking about "girl power" but I still would not let myself believe that I was to be the Young Scientist of the Year for 1999, winner of IR£1000 and Ireland's representative at the European Union Contest for Young Scientists at Thessaloníki, Greece, in September. It was just too scary.

The stage seemed enormous as I walked out to be presented by the Taoiseach with a magnificent trophy. I remember talking to him for a while, and afterwards there were lots of photographs taken. The rest is just a blur.

Part IV

After-Math

13 Media Blitz

In the rush of photographs, interviews and handshakes that followed backstage after receiving the award, I don't believe I got one moment to myself to step back from it all and let what had just happened sink in. "Sarah, are you delighted with your win?" But of course I'm delighted—I just don't feel delighted yet. When I eventually got past the wall of attention after interviews with Radio Television Eireann Television News and International News Network, I was met by Mom's smiling face and those of my brothers Michael and Brian. At that stage what I would have most liked was to talk to Mom and Dad about the events of the last few hours. However, Dad had remained at home with my two younger brothers, David and Eamonn, but even if he had been there I could hardly have unwound just then as my voice was almost gone because of all the talking.

The *18:01 Bulletin* and later the nine o'clock news that Friday night reported on the competition in general. They both showed a clip describing my project and its possible applications. This was followed by a short interview with me and then one with a judge, who commented, "If she plays her cards right I think she should make a lot of money. . . . "

Saturday morning's papers gave a lot of coverage to the Young Scientist Exhibition with photographs of some of the contestants and descriptions of projects. As overall winner I certainly got a fair share of publicity, with photographs of me appearing on the front pages of some of the national papers. The *Irish Times* carried the article with the headline JUDGES DESCRIBE WINNER'S PROJECT AS "BRILLIANT" and a picture of Mom and me underneath. Other papers also had very flattering headlines along with paragraphs discussing the project. One, under a section heading "Huge Implications," read,

> She tackled the encryption of messages as used by banks, business and governments to safeguard information. The security of these messages is vitally important for business, particularly where credit card and smartcard transactions are concerned. Her new computer system could have huge implications for the industry.

My display had been moved from its stand up to the main stage immediately after the award ceremony, and for the entire weekend the stage was absolutely swarming with people wanting to see "the winning project." Many stayed for a long time reading it, some even took notes. Although I wondered what they made of it all, I spent most of the time out of sight as I was shattered.

That night in the hotel I was to meet all my Dublin relations —my three uncles with their families and my grandparents—and of course Mom, Michael and Brian. I remember asking Mom whether or not Dad might come to Dublin that day or the day after. I was really disappointed if not a little annoyed when she said she didn't think he would. Later, when I saw him walking in with my two younger brothers to the room where we were all gathered, I rushed over to him, hugged him and said, on the point of tears, "I'm so glad you're here, I've had nobody to talk to." After drawing him to one side I gushed technical talk for at least a quarter of an hour; what I couldn't discuss with others I now poured out. I told him about how thrilled I was to have been able to explain such intricate things to professionals and nonprofessionals alike, but more than anything how great it felt to see my understanding and explanation stand up to the very searching scrutiny I got at every turn. As I spoke I could feel the tension leave my body and relief slowly take over. Only later in the bathroom when I burst into tears of happiness did I realize that all the praise and admiration meant nothing until I had spoken about the real reasons I had won the competition. After that private conversation I really began to enjoy myself, and I had a great time joshing with my sharp-witted uncles, who were determined not to let me get a big head.

One very funny thing happened just as we all stood up to say good night. When I went over to Dad I surprised myself so much by what I did that I just broke down laughing, along with everyone else who saw me: I unconsciously stuck out my hand to be shaken. Dad gave me a puzzled look, but it is only when he took my hand that I realized what I had done. I still cannot believe that I did that. I must have shaken a hundred people's hands in the last day and a half and was just programmed by that point. I turned off automatic pilot and gave Dad a proper goodnight hug, but I shall never forget that moment as a reminder of just how mentally tired it had all made me.

The following morning, Sunday, the major award winners met President Mary McAleese at her official residence, Áras an Uachtaráin. I was allowed to bring my four grandparents along with Mom. Dad's mother had an irrepressible smile on her face for the entire visit, and when the President was introduced to my mother's father he brought the conversation around, as only he can do, to a discussion of the trophy I had been given. He informed her that while its silver top in the shape of a tetrahedron had three different scientific themes carved into its faces, its base is made from 4800-year-old Irish bog yew—something which we all agreed was humbling and which makes it very special.

Later that afternoon, back at the exhibition hall, I couldn't help smiling on overhearing someone say, "And to think that people spend thousands of pounds to send their children to posh schools!" I knew that Vincent and I had to be in some way the subjects of this conversation, and I felt pleased for our teachers that this was one of the reactions to our school's double victory.

I must admit that I was a little uneasy about going back to school because I was unsure what the reaction of my friends and fellow students might be. Would they suddenly not want to talk to me because they thought me a swot?* If so, that was their problem. But I was just being paranoid. When Vincent and I arrived back for classes our fellow pupils and teachers were proud and

*In American, that would be someone who studies a lot.

thrilled, and regarded us as two heroes returning to their school. One of the nicest things anyone said to me, strangely enough, was a simple but sincere "Fair play to ya, girl" (actually it was "Fair f—ks to ya, girl!" but anyway). Later in the week village pride peaked when the media reported on the Taoiseach's visit to Blarney to congratulate the school for its achievements and for its Young Scientist double. The importance of this unprecedented occasion really rubbed off on everyone and gave the whole place a lift. It was a very memorable day.

This was the first of a number of events at which I was to be made aware of the appreciation that Cork people show when one of their own makes them proud. About a month after the Young Scientist exhibition I was invited to a gala dinner by the organizers of the information technology exhibition, IT@Cork. I thought I was there only for the photographs, but they had a surprise in store for me. In the main speech they spoke of the aims of their exhibition, one being to publicize the development of IT in Cork. To that end they announced that they wished to present me with a laptop as a way of expressing their pride in my putting Cork on the IT map. A laptop! That is the biggest material present I have ever received.

Later I was made a Cork Person of the Month, and later still I was given a civic reception by the Lord Mayor, which makes me very proud to be a Cork person.

The day before the Taoiseach's visit an article had appeared on the front page of the London *Times* which (as mentioned already in the Foreword) carried a large photograph of me in front of a blackboard and a caption underneath which read:

> Sarah Flannery, 16, who baffled the judges with her grasp of cryptography. They described her work as "brilliant."

This article changed my life. Had it been placed somewhere in the inside pages of the paper as originally intended, I believe you would not be reading this book now. Reuters picked up on the story, and my name, along with a summary of what I was supposed to have achieved, was published in newspapers all around the world. This caption, while flattering, left me completely flabbergasted. I could not believe that a school project could evoke such interest, and I could barely take in the fact that the article was about me and not some other person.

Within hours of its first appearance press, radio, TV, Internet surfers, patent lawyers and well-meaning individuals assailed our home by phone, e-mail and letter. On that day alone we received sixty phone calls! For weeks afterwards Dad had to remain at work late into the evening to answer the flood of e-mails inquiring about my project. This was just the beginning. Our old farmhouse has a lane leading up to it so long and bumpy that we are rarely visited, even by the most eager canvassing politicians. However, this border road between our privacy and the outside world proved no obstacle to the stream of enthusiastic reporters from all over, who were to brave its horrors over the next few weeks to ask me questions.

It was unavoidable that William and Michael, too, would be sought out to express their opinion of me and/or what I was supposed to have achieved. As soon as the *Times* story broke, William received the following e-mail from an English colleague:

> From the front page of today's Times <www.thetimes.co.uk>. As usual (technically) content free. Anyone know any technical details?

The opening lines of William's reply were:

> Yes, I do. It's based on work that Sarah did in Baltimore when she was here on a student work placement last March.

These were followed by some explanatory technical paragraphs. He finished by writing:

> I haven't had time to look at Sarah's project in great

> detail so I don't know how far (or even whether) she's taken it beyond where we had it.
>
> Sarah, by the way, is level-headed enough to know that new public key algorithms only made you millions if you invented them in the Seventies. Her real problem is trying to stop the journalists talking up the stupid parts of the story while still emphasizing that there's a real story in there.

Later in the evening he wrote:

> In fact, Sarah made substantial contributions to the development of the algorithm, finding a way of trading off key generation time against encryption time and doing a lot of work on the proof of security. It's very impressive.

The *Guardian* newspaper published an article headed TALES FROM THE ENCRYPT and quoted Michael as saying the following:

> And this lassie Sarah Flannery came to do work experience from school. It was immediately perceived that she could do more than the tea, and they said can you programe [*sic*]? and when she said yes, they said: "Why don't you try and programe [*sic*] this idea of Michael Purser's?" So that is what she did. She did it in three days, which was remarkable.
>
> She's a charmer: she wins the hearts of everybody who meets her, she's slightly shy and very modest and very concerned to give credit to other people, and she has a lovely smile and long hair and she's anything but brash. But she's very sharp. If you say something to her, she doesn't accept it like that, she analyses it. She has a very good brain and a determination to understand, to make sense of what she is doing.

Knowing from my own experience that what is printed is often not exactly what was said, I was nevertheless delighted. How could I not be?

At the end of the week of the *Times* article the phone calls were still coming. Friday night, as I was about to leave for the annual athletics club dinner (which I was really looking forward to), the phone rang again. How I resisted the temptation to let it ring and just walk out the door I do not know. It was phone call number 147 of the week, and I was in no mood to talk to anyone except my athletics buddies over a really good meal. "Hello, may I speak to Ms. Sarah Flannery, please?" My "Speaking" was answered by: "This is Ron Rivest calling. I wonder if I may have a few moments of your time to speak with you?"

I couldn't believe my ears. All I could muster was, "Excuse me?" At first I thought it was Dad fooling around. It could have been, but it wasn't. I couldn't believe I was speaking to Professor Rivest from MIT, the *R* in the RSA. I was very nervous and felt so humble speaking to him. At the same time I didn't want to sound like someone who couldn't stand her ground. He wanted to talk mathematics and I was delighted to do so, no matter how uneasy the experience was making me. It was so satisfying to be asked questions by someone who would not expect trivialized answers. Having such a conversation after days of being asked to spell "algorithm" was very fulfilling. Ronald Rivest—wow!

In an interview later, when asked about me, he kindly said, "She knows her number theory." This is high praise from a man who was instrumental in developing one of the first practical applications of number theory. A remark like this, though to my mind too great a compliment, is a lot more gratifying than being described as a mathematical genius by someone who could not appreciate what that really means.

After the Ron Rivest phone call I went to the dinner and had an enjoyable time. The food and the company were great—they even presented me with a huge cake. But I couldn't wait to share my recent phone call with my parents, who were in England at an aunt's fortieth birthday party. The next morning, when I contacted them, Dad could not believe who had phoned. My uncle, having heard the shouts during our conversation, asked why Dad was so excited. Upon being told, he commented, "I take it then, that God has been in touch."

Patent Nonsense

One issue that sent a lot of people into a tizzy was ignited by the sentence

> She is considering publishing her findings rather than patenting as she does not want people to pay for her discovery.

This appeared in the second-to-last paragraph of the infamous *Times* article, which appeared under the headline TEENAGER CRACKS E-MAIL CODE.

Almost everyone asked me, "Are you going to patent it?" And almost everyone had an opinion that they were anxious to share with me, whether I wished it or not. Judging from some of the Internet sites, the issue sparked off a war of words. The following extracts are a very small sample of the many views expressed:

> God Bless this kid for not patenting her discovery.

> Yeah yeah . . . wonderful things, she's already won one for the young people, etc. etc. . . . Newsbreak for little Sarah—none of that crap pays the bills. If she's reading this then the message is simple—patent this! You're a genius and you deserve to be paid for it.

> A very intelligent but naive girl. She should patent the information and license it. If she finds the rest of her life

as productive as her teens, she could let it go for as little as she feels necessary.

I hope Sarah's parents are wise enough to talk her into a patent. While her attitude is an applaudable one, it would behoove her to retain the right to such a potentially profitable technical development. What they haven't taught young Sarah in science class just yet is that most scientific discoveries don't get patented because they have no profit potential, not because scientists are more altruistic than anyone else as a group.

Don't ever underestimate Sarah's contribution to the world. I am not only talking about the faster data transfer; I am also talking about the generosity and her down-to-earth attitude. Just imagine how much we have been paying for the patent royalties; some of them are reasonable; some of them not. Of course, one may argue, without patent protection, no one would spend anything on R and D. True. But hey, Sarah deserves our praise for her generosity.

The most important thing to come out of this, besides faster encryption, is the message it sends to young people, and especially young women, around the world. . . . Those who worry that she might "lose" because she has chosen not to patent her algorithm are missing the point, which is that she has already "won."

Well done—good to see somebody not in it for a fast buck!!

Make money Sarah! Patent it! Don't throw away millions of dollars!!!!

It is very telling to see from whence came the criticism of this young lady's decision not to patent her discovery.

The American Dream is to invent something, live off its profits, and never think or work another day in one's life. That is why Bill Gates is such an American hero. It is presumptuous and arrogant to think that this girl who has achieved so much in so little time does not understand the "ways of business." . . . Also, her fame is greatly enhanced by making her breakthrough freely available. I congratulate Sarah on her achievement! She is very cool.

Great achievement, if it is proved to be of both encryption strength and high speed. Her idea of not patenting it reminds me of the essence of science, explore the world and share the discovery. Cool.

Her act is in the true spirit of the Internet.

Sarah—Remember a woman's prerogative—you can change your mind.

Patent it. Okay? Cool.

Software patents are extremely short-sighted. Imagine if great thinkers such as Sir Isaac Newton or Albert Einstein had been more concerned about protecting their IP [intellectual property] than advancing the collective body of knowledge of humanity. Every textbook that talked about the fundamental laws of motion or general relativity would be that much more expensive because of the licensing fee. I'll give you an example: how many people outside of the sphere of mathematics have heard of the Kalmann filter? This is one of the most significant breakthroughs in recent history, yet it has not achieved the degree of awareness of, say, the Theory of General Relativity. I argue that this is due in part to the fact that it was patented.

Before the competition I had never given any serious consideration to the commercial potential of my project. We certainly discussed the issue of patenting, but Dad's strong feeling was that mathematical ideas ought to be freely shared by all. To do otherwise would be an insult to the memory of the great mathematicians of the past, who carefully recorded their ideas so that future generations could build on them. So I gave the matter no further thought and just concentrated on my work.

When Michael visited the Young Scientist hall on the final Sunday afternoon of the exhibition, he handed me a letter in which he set out his thoughts on the patent issue. Kind as ever, he had written that should I wish to patent my project, as some had urged me to do, then it was a decision for me and my parents and he would undertake to make no objections. In the letter he advised strongly against trying to patent, believing, like many, that it would be immoral.

> Where would we be if Lanczos had patented his method for solving linear equations with sparse matrices? If Pollard had patented his number field sieve? If Box and Jenkins patented their linear predictor? Mathematicians should be above such mercenary nonsense!

He also questioned whether or not the algorithm could be patented. It had been on display at the exhibition, which had been open, if briefly, to the public. He pointed out that there are many free encryption algorithms, so why would someone pay me rather than use one of these? Furthermore, to make a commercial success of the algorithm I'd need to market it worldwide and submit it to standardization bodies (more time and expense), with very uncertain prospects of any eventual payback.

Since my project had been on public display, all this do/don't patent talk was a load of stable-door locking after the horse had bolted. Had I stressed this last point more in interviews I might not have found myself on the "altruistic high ground" where the media placed me. Michael's and Dad's views, and my agreement with them, were not the only reasons I gave for not patenting. However, the press chose to emphasize my more high-minded comments and often omitted the more practical reasons.

Many, therefore, found my refusal to cash in on the overnight fortune they were certain I could earn utterly incomprehensible.

WHO WANTS TO BE A MILLIONAIRE . . . I DON'T
Sarah spurns a fortune

This newspaper front-page headline and caption, and ones like them, were a pain.

I didn't like the idea of appearing a fool to all the people who "wouldn't say no" to an inward cash-flow boost (actually I'm a loyal member of this club), and I didn't want my parents to be viewed as not protecting their daughter's best interests. All I could do was laugh when I heard, "Are you going to become a billionaire?" or "I've heard rumors at work that you're worth 3.4 million already" or "Is it true that you've signed a deal worth 23 million pounds with some technology company?" I wish. I wish. I wish.

Had I wanted to patent I would have no shortage of advice or financial backing. I got many letters from patent lawyers like this one:

> We would be prepared to handle the patenting of the encryption method in return for a negotiable fixed percentage of royalties received from the licensing of those patents. . . . Ownership of all patent rights would remain with you at all times, and our firm would have no proprietary interest in the intellectual property rights.

This same firm, offering unsolicited "venture capital advice," said in an e-mail to my father that they were "horrified to see that she is considering publishing it without staking a claim to her intellectual property."

Within a matter of weeks, patenting the CP algorithm would be ruled out even more strongly, though it remained an issue in the public mind for many months. However, this is a story which must wait until I deal briefly with another issue of "perception" which for me was more worrying.

Actions and Reactions

My parents had mixed feelings about everything that happened to me after the Young Scientist competition. They were, of course, delighted with my win but had grave concerns about many things, in particular the "genius" label. They were afraid of the consequences for me of an exaggerated assessment of my intelligence being accepted as fact, and were anxious that I should not be blown out of proportion, as this could only be a burden on me. They warned me continually to stick to the truth and not to allow myself to be manipulated so that I'd be perceived as someone other than who I am. "Don't let people make assumptions about you" is a statement I was later to hear from an IBM executive in Milan, and that pretty well hits the nail on the head.

I have no doubt that I am not a genius. I am not being falsely modest. Through my father's classes I have seen examples of true genius, and I know that I do not possess that "insight" that distinguishes geniuses from those regarded as merely intelligent. I also know of their complete dedication to and obsession with their interests. But they possess that extra vital 1 per cent in Thomas Edison's quote, "Genius is one per cent inspiration and ninety-nine per cent perspiration." I was fortunate to be interested in my project topic. Although I was interested—very interested—and worked very hard, I would never say that I was obsessed to the point of losing sleep.

It is not always easy to keep a grip on reality when people are assuring you that what you have done is out of the ordinary and that it merits the description of genius. You even grow tired of telling them to calm down and not go overboard. What's more, you wonder if they think you are playacting by making a pretense at modesty. Deep down, though, you know that all this genius talk is a load of old blarney. I don't constantly dream of mathematics or cryptography to the exclusion of all else; I'm as often thinking about when I can next arrange to visit my aunt and her horses, or I'm on the phone to hear what my friends are up to.

How did my brothers react to all the fuss? The two older boys, Michael and Brian, were consistently coy. On the day I won the Young Scientist award I remember asking Mom, "Where are the boys?" She told me they were hiding behind poles so as to avoid being caught by a photographer. Later they grew to resent any media intrusion into their lives, and disappeared whenever necessary. On one occasion when Dad asked Michael to join a family shot, exhorting him to "Do it for Sarah," Michael replied, "Dad, this isn't for Sarah, this is for them." Dad let him be, asking only that the next time he took a "stand" on an issue, it be a little more serious in nature. I don't want to make Michael out to be a grouch—he's far from it. He doesn't mind my doing interviews or photos and he's interested to see articles about me; he just doesn't see what he has to do with anything. That's OK. When he's a famous rock star I won't get into a photo with him. He made sure that all the attention didn't go to my head, without, of course, going over his daily limit of ten words. "You're seriously overrated, you are." The extra "you are" was for emphasis. Brian liked to appear indifferent, but always wanted to know what was happening. His main concern about all the attention and publicity was that we'd all be seen as getting above ourselves. With his usual frankness he allowed himself to remark, "People will think we're posh gits."* Charming!

My two younger brothers, David and Eamonn, didn't let any macho egos get in the way of telling their "big sister" how proud they were of her. They never resented any of the attention that surrounded us all for a while. In fact, two more forthcoming interviewees were never interviewed, nor were more willing subjects ever photographed. They loved all the excitement and insisted on helping the TV cameramen by carrying their very expensive equipment.

*Translation: snobby jerks.

The CBS crew were so taken by something Eamonn said about the persistent "buuurrp, buuurrp . . . buuurrp, buuurrp" of the telephone that they included him in their piece, which is something that made him very proud.

It wasn't always fun for the "two smallies." There were times when Mom, Dad and I were so busy that the younger boys didn't get their usual quota of attention and were lucky to be read their bedtime story. More often than not they were left in the care of their two older brothers, for whom the basic rule of society is that "might is right." Yet they never complained, and took it all in their stride.

About a month after my Young Scientist win I was approached by an advertising agency representing Pepsi-Cola to do a billboard advertisement. The amount of money being offered would keep me going for a whole year in university, all for a few hours' "work" in a photo shoot, so when Mom and Dad advised me strongly against doing the ad I felt frustrated. Their seeming unwillingness to make a penny out of my success contrasted with the way newspaper and magazine companies and photographers were benefiting from my fame. I had never asked any one of them for money, nor was any ever offered.

Admittedly, I wasn't keen on seeing my face plastered on billboards, but it had already been all over the newspapers, so that wasn't going to be an objection. "It's innocent," I argued, "it's an ad for a soft drink, not cocaine!" My parents weren't convinced. They were afraid that I would be seen as a Pepsi girl rather than as a young scientist, and that my achievement might be trivialized as a result. Maybe. They might have considered an ad with a health theme along the lines "it's not smart to do drugs or smoke," but their fundamental objection to the proposed ad which I found most telling would be the big lie that I drink Pepsi exclusively. Even if the ad didn't quite imply this, it would still be a lie for me to claim that I think Pepsi the best of all the cola drinks, and even if the ad didn't say this either, it might suggest, however indirectly, that Pepsi makes you clever enough to win science contests. This would obviously be a lie, and though I know it's not the same thing as saying "clever people drink Pepsi," advertisers know that we don't always distinguish between a statement and its converse. I wouldn't mind doing an ad for something I know about and believe to be worthwhile. I would have no hesitation, for instance (and I wasn't asked to say this), in recommending *Mathematica* because I have used it many times and find it wonderful. I'd be prepared to say so and wouldn't refuse an offer of pay for it.

Another well-known "company" who were very much taken by my success were none other than the Spice Girls. Their promotion company phoned Mom one day seeking permission to include me in their magazine. Mom thought this was charming and a bit of fun, and said yes. Dad's students, who had been very politely deferential about my Young Scientist success, abandoned all reticence and were full of interest and questions when they read "Smart Spice—Pop Babes Think Irish Whizzkid Has Girl Power." The males in particular wanted to know if they could be "fixed up"—with a Spice Girl, of course, not me!

Among the many communications I received was a series of four-to-five-page handwritten letters which, when deciphered, proved to be proposals to set up a company. Typically they detailed everything about the new company the author and I would form together. The advantages and disadvantages were laid down clearly, though naturally there never were any real disadvantages to speak of! Sources of capital were no problem. I could get back to my schoolwork, equity funds would be set up, my parents could be directors, and so on. The wackiest letter I got contained, along with other curious statements, the following:

> We can split the codes stocks. I want to keep 53% (I was born in 1953) because you are a minor and your parents can (if they want) be co-vice presidents/co-consultants in the new code company. . . .

Mom replied very delicately to this letter and didn't mention what I would have been unable to resist pointing out—that I was born in 1982!

Only once did I meet someone in person who wished to discuss the possibility of forming a cryptography company. About two days after the *Times* article had appeared I was in Dad's office in CIT surfing the net for papers while waiting for him to return from a meeting. The phone rang, and the caller, speaking with an American accent, very slowly asked if he could speak with David Flannery. I explained that he was not available and suggested that I might do. (Every phone call Dad had that week was from somebody inquiring about me.)

When Dad returned to the office just minutes later, I—excited as ever—told him of the phone call I had just had. The caller had said that he was phoning from London in the hope that he might meet both my parents and me in a few hours. He elaborated by saying that he was a businessman who formed companies based on intellectual ideas, and he mentioned the names of one or two which sounded familiar but didn't register with me that strongly. I was preoccupied with wondering how he saw himself getting to Ireland, never mind Cork, in such a short time. When he went on to say that he had a private jet and that he could meet us later that evening at some hotel of our choosing, I was bowled over. Recovering some composure, I asked him to phone our home in an hour.

The conversation Mom, Dad and I had at the kitchen table that evening about the impending visit of one so rich that he didn't depend on commercial airlines was unlike any other we had had in the past. We decided that it could do no harm to meet him; if nothing else it would be fun and it was sure to be informative. We even wondered if it was all a spoof, but when he phoned again as promised we made an appointment to meet him at seven-thirty in the Blarney Park Hotel three kilometers or so from our home.

As Dad and I sat in the foyer waiting for the arrival of our visitor, we wondered as each male entered through the glass door, "Is this he?" Will he be on his own or will he be accompanied by a group of business types in smart suits? Dad, of course, was his usual casual self in regard to dress—just about passable but with his hair barely combed (which Mom says is far worse than not combed at all).

The American arrived on his own by taxi—a short, white-haired man impeccably dressed and close to sixty years of age. After introducing ourselves we made our way into the restaurant, which was very quiet, and seated ourselves as far as possible from the few other diners so that we could speak in relative privacy. The American ordered a fish plate, while Dad and I had coffee, as we had just eaten. Our visitor began by telling us a little about himself as a businessman and then as a family man. Although he had left us in no doubt that he was fabulously wealthy, he was at pains to point out that one of his sons, after trying a career in business, had found his true vocation as a research worker and was now happy living on what he earned. He mentioned that all his children had been educated at home before going to university. That was something I knew would impress Dad a lot—let's just say Dad has very unorthodox views on education. Our guest then inquired a little about us.

It was at least twenty minutes before he started to talk about the purpose of his visit. He explained again that he seeks out inventors, intellectuals with promising ideas, with a view to setting up companies around them. Those with the ideas are made major shareholders but are very much left to continue doing what they do best, while people skilled in management do the day-to-day running of the company. When Dad asked if he could be described as a venture capitalist, he became, I think, a little offended. He was keenly interested in seeing ideas reach their commercial potential and saw himself as much more than someone who just invested money to make more money. Needless to say, he was interested in the CP algorithm. He had read about it offering just as much security as existing standards and that, being twenty times as fast, it was set to revolutionize private communication on the Internet.

Until two weeks before this meeting the CP algorithm was just part of a school project. Even if I had reported on its security and speed, I had never thought about exploiting it in a commercial way. As I said in answer to a question about patenting, "Why would someone pay me for an algorithm when there are so many freely available?" Even now, with all the uncritical praise for the algorithm and its potential, I had never seen it as entirely belonging to me, though I must admit that at that very moment I would have dearly loved to be responsible for it from beginning to end, and to be absolutely sure that it was as secure as my investigations claimed it to be.

But I wasn't solely responsible for the algorithm. My project is based on an idea of Michael's that I encountered at Baltimore Technologies, so should it have commercial potential, Michael and Baltimore would have to be in on it. In fact, because of all the hype surrounding the algorithm I was due to visit Baltimore the following weekend to discuss with them how best to handle all the publicity we were both receiving. Our visitor asked us a few questions about the company and I explained my connection. It was clear that he had done his homework because he knew that they were in the process of merging with another cryptography firm.

Although I didn't want to chase our visitor away without hearing what he had to say, and without getting some glimpse into the world of high finance, I felt I had to point out to him that the algorithm which was the reason for his visit had not, at this early stage, received any extensive form of peer review. I explained that such a process must be undergone before any algorithm is established as being satisfactorily secure; and that, while the judges had said in their citation "the demonstration of why the code is secure is publishable" and the external mathematician had been very impressed with the mathematics, neither he nor the judges, as far as I knew, were cryptographers. I also told him that Baltimore Technologies were much more circumspect in their statements about the algorithm than the media because as experienced cryptographers they know that the wayside is littered with countless failed cryptosystems which once held promise. As a reputable company Baltimore didn't wish to pronounce on the CP algorithm without giving it a thorough examination, now that it was receiving such attention and being associated indirectly with them.

I don't know what he made of all of this candor. It didn't appear to upset him in any way, though I wondered whether he was beginning to think his whole journey a wasted one. I told him that

I had made all these cautionary comments to reporters, but the articles they then wrote omitted them for fear, I speculated, that they put a dampener on a wonderful story.

He listened very attentively, and spoke in very general terms about how one would form a cryptography company, while mentioning sums of money that seemed to be in multiples of billions rather than millions. After about forty minutes I noticed that he was starting to begin sentences with "We." This was scary. It made me feel as if I had somehow committed myself to something when I hadn't. At one point he casually asked what the combined worth of Baltimore and their prospective partners might be. "About $700 million?" he speculated, and continued by saying that it would be only a matter of "our" buying them out. Whew! He wasn't being boastful, just matter of fact. When Dad said that I owed everything to Dr. Purser and couldn't dream of talking about commercial ventures without first seeking his opinion, he remarked, "We'd look after the good doctor." I didn't quite know what to make of this, but it didn't sit well with me. Dad nearly had a heart attack when our visitor asked to accompany us on our visit to Baltimore. As politely as he could, Dad said that though his suggestion might seem very businesslike, it was completely out of the question for a number of reasons, most of which had to do with good manners and friendship. Can you imagine what it would have looked like if I had arrived at Baltimore with a hard-nosed American businessman acting as though he were protecting my interests? It was a relief to talk him out of such a crazy notion without being rude. But he was very interested in exploring the possibility of exploiting the algorithm. We left him in no doubt that if such a thing were ever to take place it would have to be with Baltimore, and parted with a promise to phone him after our weekend visit.

The conversation had lasted well over an hour. As we were all leaving the hotel I was surprised to see him approach a local man as if he knew him. He did. It was the taxi driver who had brought him to Blarney waiting to take him back to Cork Airport and his private jet, which was on standby to take him to Nice in the south of France. Wow!

On the way home we talked a little about the problem of selling somebody a secret. How do you sell a secret to someone without that person getting to know it before he has paid you, and how does the person who is buying it know that he is getting the real thing?

At the end of the week I visited Baltimore, and was warmly received and congratulated on my Young Scientist win. More cake was had by all. Afterwards William and I spoke about the issues that arose from the unexpected publicity that the company and I had received. William kindly remarked that Baltimore had got ten times more publicity through their association with me than they had received some months previously through their involvement with the first "digital signing" of an agreement between President Bill Clinton of the USA and our Taoiseach, Mr. Ahern. He also joked that each mention of Baltimore on the front page of a newspaper was worth about IR£30,000.* I'll have to send them a bill.

We discussed examining the CP algorithm in more detail now that the world had taken more than a passing interest. We thought perhaps that I and Baltimore should (when we both had time) prepare a paper with a view to publishing it in a journal or, failing that, post my project report on a Web page with a commentary based on our subsequent findings. There was no talk of making millions, and as far as they were concerned I was free to talk to whomever I liked about whatever I liked, write whatever books I wished, provided I promised to say nice things about them, and go into business with whomever I liked.

Almost immediately after the meeting I phoned the mobile number I had been given by the American visitor, but I couldn't get through. I checked with directory inquiries, national and international, but to no avail. The following Wednesday he phoned and I read him back the number I had taken down. I had been missing a digit. I apologized for appearing indifferent but told him of my attempts to contact him. We spoke for a few minutes about all sorts of possibilities. He was going to "get back" to me, but as

*At the time, over US$40,000.

yet I have heard nothing. Was he for real?

By the end of January I had spoken with over three hundred journalists from newspapers, magazines, Web sites and TV. I never refused to do an interview—I spoke to everyone. Not all the phone calls I received were from journalists and some I would not have wished to miss, like those with invitations to take part in events in places as far away as Singapore.

I also did many radio interviews for stations such as Radio Colombia in South America (that was an interesting one—parallel translation), and Greek, French, German, American, South African, English and, of course, Irish stations. Those interviews *as Gaeilge* (in Irish) put me on the spot. My Irish can't be described as fluent and I knew my teacher would be listening in, ready to take down all my mistakes for the next class. Interviews for magazines were usually the longest and often led to two- or three-page spreads, some of which I never saw. The interviewers were usually well prepared with questions and lots of clippings for me to comment on. One German reporter, on hearing that Dad and I were off to the Mathematical Excursions evening class, asked if he could come along. He wanted to witness Dad's lecturing and experience the atmosphere. He sat next to me for one and a half hours and seemed to take a keen interest. He must have enjoyed his trip back to school because he wrote a glowing piece for his Hamburg newspaper.

Because I am self-conscious when it comes to any kind of camera, I hated having to appear on television. I found such appearances so tiring and stressful that I swore "Never again!" after I had done three or four of them. An Irish and a German TV crew filmed at home and at school for one and a half days each. Filming at school was the worst. "Just ignore us." Yeah! Right! Whatever! The crew would walk behind me into classes and follow me through the corridors with their big black cameras. I worried what my classmates or others in the school might think, but some lapped up the attention. A team from CBS News's *48 Hours* program filmed me for three days at home, in the Mathematical Excursions class at CIT, at school, horseback riding in Blarney Castle grounds, playing basketball, even eating my dinner. Although they were extremely considerate, very soon it all got too much for me and I just wanted to run away and hide.

I received many invitations from universities and colleges around the country to give lectures on the CP algorithm. These talks were meant to be specific in nature, and were presented at math department seminars or meetings of student mathematical and computer societies. At the time I was so snowed under that Dad kicked for touch* on my behalf on most of these flattering offers. However, I received two invitations which I simply couldn't refuse, and a third which I accepted for reasons I'm still not sure of, but which I think had something to do with seeing if I had enough nerve.

*"Kick for touch" comes from rugby, but in everyday talk it means asking for time before giving a definite answer to a request.

14 Around the World and Back

Very early on I received three invitations I felt I couldn't refuse. Two were to other countries, but the scariest one was closest to home. The first was to Singapore. I and one of my parents were to be guests of the government for an all-expenses-paid trip. In return, I was to give a speech at the closing ceremony of their National Science Talent Search Contest and a number of talks at some of their secondary schools to groups of young people of my own age. The rest of the time (a whole week) was my own, and they'd throw in some spending money. Even if I were the shyest person in the world I wasn't going to turn this one down!

The flight from London to Singapore took thirteen hours. I was very excited to be in such an interesting place and to have a whole day to myself before giving a speech at the closing ceremony. Singapore is, as was explained to me, very much a knowledge-based community, and the government is particularly supportive of the education system. It is very impressive judging from what I saw while there, and the schools I visited were like our third-level (that is, post-secondary) institutions in terms of their facilities.

That evening I tried to rehearse a speech I had written for the following day's ceremony. As I tried to block out everything else while pacing the balcony of my sixty-third-floor room, Dad delivered frequent but most unhelpful assurances to me that I would be f-i-i-i-ne. I wanted to be very familiar with the flow of thought I had worked out and to avoid, as much as possible, reading the script of my speech—a speech which I was now beginning to really dread. What if I messed up after they had flown me all the way from Ireland? To add to the pressure I was scheduled to speak after an award-winning professor and before a very successful researcher.

On the morning of the ceremony the professor began his talk with a lighthearted joke as he moved easily around the stage. I wished he were not so good; every aspect of his speech increased my nervousness and reduced me to a quivering wreck. When his slot was coming to an end and his PowerPoint presentation was about to wrap things up beautifully for him, I, who had only been concentrating on his speech, realized with a shock that I was up next. Yes, here it comes; the master of ceremonies is calling me up onstage after making me out to be the bee's knees.

I reluctantly made my way up the steps and hid behind the nice thick wooden lectern opposite the modern see-through one on the other side of the stage. I unfolded my crumpled pages, cracked my first joke and concentrated on looking first to the left of the room and then to the right, but never to the middle where Dad was sitting. Besides avoiding looking at Dad, I was at the same time trying to block out two other rather irritating things—my whole left leg was shaking violently, thankfully in the sanctuary provided by my big blocky lectern; and my face, my quivering white face, was being projected onto two very large screens at either side of the stage.

I was concentrating so much on speaking clearly and distinctly that it was only when I slumped into my chair afterwards that I became aware that I had most certainly run overtime. Having waited every agonizing moment up to my own time to speak, above all I didn't wish to spend an eternity speaking to my audience. I also felt bad about not having made use of a PowerPoint slide presentation, though I had one almost ready. In my few rushed moments that morning I did not have time to make sure it would run smoothly, so rather than have another thing to worry about while standing in front of a very varied audience I easily persuaded myself not to bring it, much to Dad's disappointment. Being so tired I had chosen to ignore his advice and go against even my own instincts. It really would have improved the presentation.

Surprisingly, the part of my speech that everyone picked up on was my discussion of Dad's method of teaching us through puzzles on the blackboard at home. Thankfully, this took the attention away from me afterwards, and he was quizzed instead. Later I was chuffed* when one of the organizers was able to remember all the main points of my speech in perfect order—even the ones towards the end! When we finally took our leave and went back to the hotel room I was dreading Dad's few words—the inevitable "should have had the PowerPoint" came up, but apart from that the praise was flowing. But I was not accepting any, as I was hung up on the mistakes I had made. I wanted to learn from them, and I wanted to concentrate on the next few days when I would have more opportunities to make amends.

How on earth is everyone so lively here at six in the morning? I was due on the other side of the city by eight-thirty for the first of four hour-and-a-half-long talks, now thankfully reduced to three. I had the laptop with my PowerPoint displays this time, and was guarding it dearly on the way to the junior college that was to be my base for the day. These talks were to be more general than the speech at the award ceremony and therefore not so difficult. I had some nice presentations, different ones for different levels of audience. I was still very nervous, as I would be talking first to people of my age and then to older students for the remainder of the day. This was rather daunting, especially considering some of the projects I had seen at their science competition. I spoke to each group about why I was there, and how I had got interested in math and cryptography, and I showed them a few puzzles before taking them in detail through the one on magic squares. I used this to explain that "a little thinking can save a lot of computing," and how by exploring just a bit beyond what you were initially asked you could end up proving something (in this case exactly how many magic squares there are). To finish up I went through some of the main ideas of cryptography. There was half an hour allotted at the end of each talk for questions and answers. If this dragged, despite teachers' efforts to get their students to ask more questions, I would throw out a few more challenging puz-

*For readers who speak only American, I was happy!

zles, to which most responded with enthusiasm.

I was exhausted by the time it came to give the last talk that Monday evening, and while waiting to be asked up onstage I nearly fell asleep. However, this talk proved to be the best. Judging from the questions I had been asked after my previous two talks, I now knew which points to explain more fully. I also felt that this time around I had given just the right amount of the personal story. The teachers present were most appreciative of the work I had put in, and it was they who asked the best questions. All in all, I enjoyed the presentations and felt good about having come so well prepared. I talked without notes and so looked at my audience the whole time—a big achievement for me!

I gave one more talk on Tuesday to a more varied group of students at another school. That, too, went well. I was now free to enjoy the rest of my short stay and experience some of the culture of the East. I visited the Indian and Malayan markets in search of exotic presents, and bought some silk for Mom. In the evenings, Dad and I searched out different restaurants, where we experimented with all sorts of Asian cuisine, before heading for Boat Quay to sit out in the warm night air drinking beer while soaking up the atmosphere (which, if the truth be told, was more Western than Eastern). There was little fear of ever getting lost on our way back to the hotel, the Westin Stamford, because, being the tallest in the world, it is visible from almost every direction. It was disappointing to be returning home so soon.

After the twenty-three-hour journey home I had exactly seven hours before I was off again—not to another speaking engagement, but on a school tour of Dingle and the mountains of Kerry! I had looked forward to this weekend before ever going to secondary school, and comatose or not I was going to enjoy it. Sometimes the body needs diversion more than it needs sleep. It turned out to be the most fun I had all year. We climbed Mount Brandon and rolled back down in the snow—twice! I laughed so much my stomach hurt. The day after that Mom and I drove to Leenaun in Galway, in order to fit in with the recording schedule of Channel 9, Australia, to do another interview. They were to film in Cork the following day, but by that time I—jet-setter that I had become—would be in Milan for my next engagement.

I was also invited by IBM EMEA (Europe, Middle East and Asia) to be a guest speaker at their first ever leadership conference for women. The conference, in Milan, had the theme "Women and Technology in the New Millennium." Apparently I was wanted as a speaker because I was an example of "the highly intelligent, motivated and determined young women who are set to change the face of our industry in the years ahead." Oh! Imagine what an expert-in-everything the audience would be expecting on seeing such a description in the program. Very intimidating, but again the invitation was more than generous. My parents and I would be put up in a fancy hotel, and chauffeured to and from all events, and I would receive a fee for speaking that ranked (from my point of view) with some of the numbers used in cryptography.

Despite my experience in Singapore I was never so nervous about anything in my entire life as I was about this presentation in front of two hundred women at the IBM conference. Never! However, my talk (this time with a PowerPoint presentation) was very well received—in fact, I was given the only standing ovation of the three-day conference, though I think it might have been out of pity for the seventeen-year-old who had to speak to all those successful career women. I listened with great interest to a number of the other speakers, hearing the kind of things that occupy the minds of businesspeople—how to improve sales, how to attract a suitable workforce. On this last point one speaker said that the cleverest students, as judged by grade point average, don't necessarily prove to be the best employees—her actual words were, "Next to eye color, grade point average is the most useful indicator of a suitable applicant." They must have liked the color of my eyes because a week later I was offered a summer job—at any IBM site of my choosing.

The third invitation was to be guest speaker at the annual general meeting of the Dublin Mathematics Teachers' Association to be held at St. Patrick's College, Drumcondra, in Dublin. Speak in front of math teachers? No way! Why I didn't turn this down I'll never know, but I accepted.

This was quite a turnabout—here was I, standing in front of teachers. But unlike a class of students, they had turned up voluntarily. It was to be my biggest challenge yet because I would be addressing a mathematically educated audience who would follow what I said very attentively. I went to a lot of trouble preparing a PowerPoint presentation that included more advanced material, to lend more substance to the talk.

No sooner had I delivered my opening lines than my computer crashed. My confidence must definitely have been increasing because I surprised myself by remaining outwardly calm and rebooting the machine while continuing to talk. Had this happened to me during one of my earlier talks, I would have crashed too! The presentation went smoothly after that, and I could sense that the teachers were quite pleased. With at least one number theorist in the audience that I knew of, it was only to be expected that the questions afterwards would be tough—and they were. They were the best I had been asked in all of my talks.

The Attack

As I have said, right from the outset Baltimore Technologies were always very careful not to be drawn into saying anything exaggerated about the Cayley-Purser algorithm. In the first few days after my Young Scientist win they couldn't have made any informed comment because they hadn't studied my latest project report. Although Michael and William had visited the exhibition, they naturally spent most of their time talking with me rather than examining my display in any depth. Baltimore had to be as surprised by the sudden attention they were receiving as I had been when it first came out of the blue at me. Of course, it was my naming of the CP algorithm and my informing the press of the connection that drew all this publicity upon the company. For the first week they were inundated by journalists looking for a story, just as I was, so they had little or no time to give attention to my report. However, Michael, because he was so directly involved with it and because he shunned most interviews (leaving them to William, as he was happy to admit), began to think his way back into the ideas. He harbored misgivings about the CP's security because he had eventually abandoned his own system after finding it not to be entirely to his own satisfaction. Almost a week after that fateful *Times* article, he e-mailed to say that he recalled an "attack" on his work which had been pointed out to him by a mathematician and which he felt might also compromise the CP system. An attack on a cryptosystem is something like a prisoner making a Shawshank-type escape from a high-security prison from which escape was believed to be "impossible."

When I first heard about the possibility of an attack on the CP algorithm, I was desperately disappointed. Now that I had been in the public eye so much I couldn't help feeling that I had perpetrated some kind of giant fraud. Deep down, of course, I knew I hadn't done any such thing. I had repeatedly said to reporters that the algorithm was being praised before it had been properly appraised, and that it had not been subjected to the kind of peer review by competent cryptographers that is necessary before any system can be judged to have any merit. Although some did report these cautionary comments, many chose not to. I know most people wouldn't care a straw one way or the other, but for me it was vitally important that people in the mathematical world who had helped me and whom I respected could see that I wasn't making any over-the-top claims. I found it very comforting when many mathematicians Dad met afterwards commented on how careful I had been to avoid exaggeration.

Dad, Mom, Michael and William all knew of my discomfort and tried to put me at ease in every possible way. My parents said that it is quite normal in science for one researcher to report an investigation and for another to point out something not quite right or suggest an improvement—that science moves on as much by negative results as by positive ones. A mathematician whose work is found to be less than perfect is still regarded as an active contributor to progress. If things were otherwise, nobody would publish. Michael wrote, "Nor is your achievement any less because there is a security problem with the algorithm. There are very few algorithms which stand up to rigorous analysis."

I knew all this was true, but because I had been made out to be such a hotshot, would the papers suddenly cry "Fraud" if I were to reveal the attack? One newspaper had run an article with the heading NO MILLIONS FOR SARAH. Although it was probably the first hint of a negative note in what so far had been presented as a fairy story, I wasn't upset by it. It was more realistic than most reports up to that time. However, I had no idea that the article was to appear, and it was chilling to realize that reporters could write almost anything they wanted without bothering to check with you on points of fact or how you might feel. One or two articles had been very cavalier about what most would regard as trivial details, such as inventing a description of our kitchen (I'm still looking for the pine table we're supposed to have) and quoting me as saying "girly" things that I would never say in a million years. So I was uneasy because I felt that if knowledge of the attack became public prematurely, there was no telling how the press might react. I was also worried for another, more serious reason.

Dad remarked in an e-mail to William that it was ironic that it is the Cayley-Hamilton theorem that undermines the Cayley-Purser algorithm as a public key system. By way of reply, William commented, "Either way, Cayley wins." His humorous observation highlighted something that I hadn't thought about up to now. When dealing with the first burst of publicity I was afraid of drawing attention to people whom I knew had no interest in receiving it. My gravest concern had been for Michael. By titling my project as I had, to acknowledge how much his ideas had helped me, I had unwittingly brought the attention of the press upon him. I know him to be a man who owns no television set and refuses all TV interviews, so you can imagine how much he values his privacy. (I am conscious that by writing all this I am only further adding to his exposure.) But now the attack was an even greater worry to me because I feared it might affect how he would be remembered.

I knew from listening to conversations between scientific friends of my parents how true scientists really despise those who publish any old rubbish to pander to the "publish or perish" maxim. (Dad jokingly said of the attack, "publish *and* perish.") I knew also from correspondence with William and Michael that they were both men who would have no truck with any kind of "soft science." They were conscious of their reputations, and of Baltimore's as a serious cryptography company. Even before the attack reared its head, Michael had written, "We [Baltimore, William and I] could look very foolish, lucky you would be excused on the grounds of youth." Months earlier I had been very much struck by William's saying, "Bad business is one thing, but bad science is another." So I was very conscious now that after a lifetime devoted to mathematics and producing work of the highest quality, Michael might be more thought of as the person connected with "that faulty algorithm."

Now that I knew of a possible attack on the CP algorithm, I had first to learn its exact nature. I felt under pressure as people were hoping that I would publish it or post it on a Web site fairly soon. Naturally, Baltimore (through Michael and William) and I wanted some time to try to find what is known as a patch. In stark contrast to those wonderful March and April months when I was full of energy and enthusiasm and had time to investigate, I was now back at school and worn out from either doing interviews in the evenings or being followed around by TV people, while all the time having to deal with letters and e-mails. As you can imagine, it was very difficult to recover the proper frame of mind necessary to do any kind of work involving clear thinking.

The attack, if it proved to be successful, essentially meant that the CP algorithm could no longer be classified as a public key system, which was its main claim to fame, though since it works as a *private* key algorithm, it could still be regarded as a valid cryptosystem. But it held little attraction for me unless it could be restored to its original status.

I took a long time to settle back to work. When I eventually did, despite exploring different variations that held out promise, it turned out that my feelings of hope were always short-lived. I learned a lot more math, but unfortunately I also learned in the end that the algorithm couldn't be patched. Its public element simply gives just a little too much information to the bad guys. When I reflect on the whole experience of working on the project, doing "real" work, with Dad, Michael and William, which showed me a whole new side of life and gave me a real glimpse into what it might be like to work on "mathematical things" in the future, I realize that I learned so much and gained so much from it that to be anything more than just a little disappointed would be silly.

Whereas Baltimore had been anxious to publish something a month or so after the storm of publicity first blew up, I asked that we not do this until I had attended the European Contest in September. Besides wanting to stay quiet about the attack for a while longer, I didn't want the status of the project as Ireland's entry to be compromised by any discussions that might arise on the Internet.

Throughout this whole period Dad, Michael and William snatched whatever time they could to think about and discuss the attack with me, as did one very renowned mathematician. On reading my report he immediately saw where the CP system was vulnerable, but thought it a matter of little consequence in the grand scheme of things. I was really cheered by his attitude. His whole outlook was completely positive. He viewed the publicity for the Young Scientist '99 winning project as a great advertisement for mathematics, and he hoped my win would help lift the image of his beloved discipline and attract back the bright young minds which for the moment are being lost to the other sciences. What a lift I got on hearing this. That I might have done a service to someone or something, never mind its exact nature, made me feel good, even proud.

As soon as I finished school, I got cracking on my forms and the report, which had to be submitted by June 1 to the EU Contest judges. The forms were easy—they just wanted straightforward facts. The worst part was scaling down the project report. The one I had written for Dublin was about fifty pages long, but now "the written project may consist of up to a maximum of 10 pages of written text. It may be accompanied by up to a further 10 pages of illustrations." Ouch!

What a lot of tedious work and tough decisions—what should go and what should stay? The most difficult part emotionally was writing a postscript detailing the nature of the attack on the CP algorithm. In accordance with this "addition," I called the project for the EU Young Scientist competition "Cryptography: An Investigation of a New Algorithm Versus the RSA." In true Irish style I sent off the forms and report by courier on the morning of June 1!

As I mentioned, a week after I returned from giving my talk in Milan, IBM invited me to spend some time at any IBM site of my choosing. So beginning the second week of July, I spent four weeks at the Smart Card Division in IBM Development Laboratories, Böblingen, which is near Stuttgart, in Germany. While I was apprehensive about what their expectations of me might be, I quickly settled into my new workplace with the support primarily of my "foster parents," Elke and Stefan. They work in the division and were exceptionally kind to me, taking me to classical music concerts in the magnificent castles of the region and, best of all, taking me hiking in the Swiss and Austrian Alps. As a student intern on work experience I was given small tasks to do which involved programming smart cards using a special version of Java. This was nice because it meant I got to learn a little about this language (on top of brushing up on my *German* language skills), and I also got to know a bit about Smart Cards.

A highlight of my stay was a visit to the IBM Research Laboratory in Zurich, Switzerland, where I was really impressed by the exciting and unpressurized atmosphere of innovation. There is little or no hierarchy in the workforce, just Dr. This and Dr. That. Each researcher appeared to be working on something unique on the cutting edge and—most importantly, as far as I'm concerned—investigating a problem of his or her own choosing. What's more, they all seemed to have good, well-paid jobs. I could work in a place like that!

The European Union Young Scientist Contest

The first day of September saw me in my final school year and signing the contract for this book. What a start to a month. Two weeks later Mom and the boys flew to Greece, and I went with Dad and a friend of his to Dublin for the All-Ireland Hurling Final (which Cork won—UP THE REBELS!). The day after, in high spirits, I paid a visit to Baltimore, where I spoke with William about the posting of the project on their Web pages once I had made it public at the European Union Young Scientist Contest in Greece. Then it was back home for a week of sustained hard work. I went through every detail over and over again, particularly the postscript, because I imagined this would be where the European judges would focus most of their attention. Dad and I ate in Chinese restaurants for the entire week (he always takes up this lifestyle when Mom goes away with the younger boys), and I watched videos at night, utterly wrecked after each day's work. I had to pass up a wonderful invitation from the Irish Computer Society to spend a week at a conference in Bunratty Castle where I could have met Whitfield Diffie, one of the founding fathers of public key cryptography. That was very disappointing, but it would simply have been too much so close to the competition.

Ευρωπαικος Διαγωνισμος για Νεους Επιστημονες

Although I got no chance to let the sun change the shade of my skin during the week in Thessaloníki, Greece, I made a lot of friends and enjoyed the magnificent food and cultural events our Greek hosts had prepared for us. All the participants got on really well together, and my gang of friends told jokes all week despite the sweltering heat of the exhibition hall—the only thing that wasn't excellent, but only because the air conditioning wasn't working.

The atmosphere at the exhibition was very relaxed, partly because, I think, the judges had been able to study the project reports since June. Each report had already been given a preliminary mark, which meant that the judging process wasn't as intense as I had found it to be in Dublin. As in all competitions I have been to, there was a great variation in the standard of presentation and in content. When it came to displays, some entrants seemed to be clueless while others had their posters made professionally and had made flyers describing their projects. This contrast in standard was probably due to the fact that each country's national contest is organized differently, some placing more emphasis on visual presentation than others.

My main judge, a mathematician, commended me on the integrity I had shown by including a detailed report on the attack. Phew! He wished to know if I was disappointed about the demise of the algorithm as a public key system. He was also very interested in hearing about the visits I had made to Singapore and Milan because he devotes much of his time to popularizing mathematics.

In all, I was visited by eleven of the fourteen judges. Some had shown my project report to cryptographer acquaintances, who, they informed me, had been very enthusiastic about it. The more mathematically knowledgeable judges asked great general questions about this and that cryptographic matter with a "What did I think of this?" and "Wasn't this cool?" All in all, I was pleased with my performance and heartened by the number of interviews I received.

The award ceremony was held in one of the summer residences of the former king of Greece under a magnificent sunset reflected in the Aegean Sea. There was a big buildup to the main prizes. The three third-prize winners were all announced before a more detailed description was given about each of the three projects that had won second prizes. When a French girl, whose stand was near mine, got one of these for her project on volcanoes, I thought I no longer had a chance as she'd had more judging interviews than I had. Dad, on the other hand, was confident that I was among the winners because after the second prizes had been announced he had noticed that we were being photographed: he knew that "there was something cooking in the pot." I didn't think so. Because of the attack I was really prepared to go home empty-handed but proud that I had done my best. I was astounded when my name was called out first as one of the first-prize winners. As I walked up the grand steps to a huge cheer from all my new friends, I was delighted. My beloved CP algorithm may not have turned out to be earth-shatteringly amazing, but it was still good science.

Moments later I was sent into an emotional spin when the president of the jury made me stay on the stage. He wished to present me with another prize, one of the honorary awards. It was a week-long trip to Stockholm in December to attend the Nobel Prize ceremonies. I couldn't believe it. I shall regard this visit as the perfect end to what has been an eventful year.

I got a celebrity welcome-home. On my arrival at Cork Airport I was met by photographers and whisked into the VIP lounge. My achievements were hailed as a great victory for Cork and an antidote to Cork's loss in the All-Ireland Football Final just minutes after the plane had landed. The first radio interview I did was with *Morning Ireland,* a news program broadcast nationwide between seven-thirty and nine. In this interview with me and "the proud parents," we commented on the attack so as to kill off all the stories about the algorithm making billions and my being a genius. With a huge weight off my shoulders, I was free at last to do what I had always wanted to do. I sent a copy of the project, warts and all, to Ron Rivest.*

I was never really sure (and I'm still not) about what I'd like to be "when I grow up." Until recently I had no definite ideas on what to study at university, other than some vague notion that it would be something in the sciences. I suppose I have been a little anxious not to be seen to follow in Dad's footsteps. However, over this last year I have decided that, to start with at least, I'd like to study

*If you're interested, have a look at Appendix A, page 271.

math and computer science. In this way I'll learn more about cryptography, which I feel will always hold a very strong attraction for me. I hope to stay at university long enough to earn a Ph.D. I trust that, in addition to being productive, my college years will be lots of fun. Ultimately I would like to be one of those lucky people who get paid for doing what they love.

About This Book

Writing this book was torture. We could never have done it without Mom's help. She was our primary proofreader, earning herself the name Mme Fowler for her readiness to cite *Modern English Usage* in defense of her proposed alterations. Later her job description broadened to include monitoring the mathematics sections for what she termed "breaches of promise"—a reference to our stated aim in the Preface that the mathematics should be intelligible to all readers. Of course, very early on we all realized that her usefulness in this role was becoming diminished, as her understanding of the material deepened with each reading to check for improvements. This led to Dad changing her title to "Gorenstein's chauffeur."

The story goes that, in the course of a long lecture tour, the famous American mathematician Daniel Gorenstein was chauffeur-driven to various venues around the USA. While Gorenstein lectured, his chauffeur sat in the back row of the auditorium. In time the chauffeur became so familiar with the material (and it was pretty deep stuff) that he joked to Gorenstein, "At this point, I bet I could give that lecture myself!" One evening Gorenstein was scheduled to speak at a small university where he guessed his face wouldn't be known. Feeling particularly tired, the master hit upon the idea of asking the chauffeur to give the lecture while he rested in the back row. The lecture went off without a hitch, and at the end the chauffeur answered without hesitation all the questions he had heard asked and answered a hundred times before—all, that is, except one last question which left him entirely at a loss. After a momentary panic, he composed himself and said, "That question is so simple, I believe even my chauffeur could answer it." The dozing "chauffeur" shamed the questioner with an immediate reply.

Mom's final contribution was the following: "Don't worry, Sarah. If people like this book, you'll get all the credit, but if they don't, Dad will get all the blame!"

Appendix A

"Cryptography—A New Algorithm Versus the RSA"

Introduction

As long as there are creatures endowed with language there will be the desire for confidential communication—messages intended for a limited audience. Governments, companies and individuals have a need to send or store information in such a way that only the intended recipient is able to read it. Generals send orders, banks send fund transfers, and individuals make purchases using credit cards. Cryptography is the study of methods to "disguise" information so that only the intended recipient can obtain knowledge of its content. Public key cryptography was first suggested in 1976 by Diffie and Hellman and a public key cryptosystem is one which has the property that someone who knows only how to encipher ("disguise") a piece of information CANNOT use the enciphering key to find the deciphering key without a prohibitively lengthy computation. This means that the information necessary to send private or secret messages, the enciphering algorithm along with the enciphering key, can be made public knowledge by submitting them to a public directory. The first public key cryptosystem, the RSA algorithm, was developed by Ronald Rivest, Adi Shamir and

Leonard Adleman at MIT in 1977. This system, described below, has stood the test of time and is today recognized as a standard of encryption worldwide.

Aim

This project investigates a possible new public key algorithm, entitled the Cayley-Purser (CP) algorithm and compares it to the celebrated RSA public key algorithm. It is hoped that the CP algorithm is

- As secure as the RSA algorithm and
- FASTER than the RSA algorithm

Firstly both algorithms are presented and why they both work is illustrated. A mathematical investigation into the security of the Cayley-Purser algorithm is discussed in the main body of the report. Some differences between the RSA and CP algorithms are then set out. Both algorithms are programmed using the mathematical package *Mathematica* and the results of an empirical run-time analysis are presented to illustrate the relative speed of the CP algorithm.

RSA Public Key Cryptosystem

The **RSA** scheme works as follows:

Start Up: [This need be done only once.]

- Generate at *random* two prime numbers p and q of 100 digits or more.
- Calculate $n = pq$ and $\phi(n) = (p-1)(q-1) = n - (p + q) + 1$.
- Generate at random a number $e < \phi(n)$ such that $(e, \phi(n)) = 1$.
- Calculate the *multiplicative inverse, d,* of e (mod $\phi(n)$) using the Euclidean algorithm.

Publish: Make public the enciphering key,

$$K_{\mathrm{E}} = <n, e>$$

Keep Secret: Conceal the deciphering key,

$$K_{\mathrm{D}} = <n, d>$$

Enciphering: The enciphering transformation is,

$$C = f(P) = P^{e} \bmod n$$

Deciphering: The deciphering transformation is,

$$P = f^{-1}(C) = C^{d} \bmod n$$

Why the deciphering works: The correctness of the deciphering algorithm is based on the following result due to Euler, which is a generalization of what is known as Fermat's Little Theorem. This result states that

$$a^{\phi(n)} = 1 \pmod{n}$$

whenever $(a, n) = 1$, where $\phi(n)$, Euler's totient function, is the number of positive integers less than n which are relatively prime to n.

When $n = p$, a prime, $\phi(n) = p - 1$, and we have Fermat's theorem:

$$a^{p-1} = 1 \pmod{p}\ ;\ (a, p) \equiv 1$$

If $p|a$ then $a^{p} \equiv a \equiv 0 \pmod{p}$, so that for any a,

$$a^{p} \equiv a \pmod{p}$$

Now since d is the multiplicative inverse of e, we have

$$ed \equiv 1 \pmod{\phi(n)} \Rightarrow ed = 1 + k*\phi(n),\ k \text{ in } \mathbf{Z}$$

Now

$$f^{-1}(f(P)) = (P^e)^d \equiv P^{ed} \pmod{n}$$

and

$$P^{ed} = P^{1 + k*\phi(n)} \pmod{n} \qquad \text{(for some integer } k\text{)}$$

Now for P with $(P, p) = 1$, we have

$$P^{p-1} \equiv 1 \pmod{p} \Rightarrow P^{k*\phi(n)+1} \equiv P \pmod{p} \text{ as } p - 1 | \phi(n)$$

This is trivially true when $P \equiv 0 \pmod{p}$, so that *for all P,* we have

$$P^{ed} = P^{1 + k*\phi(n)} \equiv P \pmod{p}$$

Arguing similarly for q, we have for all P,

$$P^{ed} = P^{1 + k*\phi(n)} \equiv P \pmod{q}$$

Since p and q are relatively prime, together these equations imply that for all P,

$$P^{ed} = P^{1+k*\phi(n)} \equiv P \pmod{n}$$

The Cayley-Purser Algorithm

Introduction

Since this algorithm uses 2×2 matrices and ideas due to Purser it is called the Cayley-Purser algorithm. The matrices used are chosen from the multiplicative group $G = GL(2, \mathbf{Z}_n)$. The modulus $n = pq$, where p and q are both primes of 100 digits or more, is made public along with certain other parameters which will be described presently. Since

$$|GL(2, \mathbf{Z}_n)| = n\phi(n)^2(p + 1)(q + 1)$$

we note that the order of G cannot be determined from a knowledge of n alone.

Plaintext message blocks are assigned numerical equivalents as in the RSA and placed four at a time in the four positions (ordered on the first index) of a 2×2 matrix. This message matrix is then transformed into a cipher matrix by the algorithm and the corresponding ciphertext is then extracted by reversing the assignment procedures used in the encipherment.

Because this algorithm uses nothing more than matrix multiplication (modulo n) and not modular exponentiation as required by the RSA it might be expected to encipher and decipher considerably faster than the RSA. This question was investigated, using the mathematical package *Mathematica,* by applying both algorithms to large bodies of text (see Tables I–IX) and it was found that the Cayley-Purser algorithm was approximately twenty-two times faster than the RSA with respect to a 200-*digit* modulus.

Needless to say if it could be shown that this algorithm is as secure as the RSA then it would recommend itself on speed grounds alone. The question of security of this algorithm is discussed after we have described it and explained why it works.

The CP Algorithm

Start-up: Procedure to be followed by Bob (the receiver):

- Generate two large primes p and q.
- Calculate the modulus $n = pq$.
- Determine C and A in $GL(2, \mathbf{Z}_n)$ such that $CA \neq AC$.
- Calculate $B = C^{-1}A^{-1}C$.
- Calculate $G = C^r$; r in $\mathbf{N}$.

Publish: The modulus n and the parameters A, B, and G

Start-up: Procedure to be followed by Alice (the sender):
In order to encipher the matrix μ corresponding to a plaintext unit for sending to Bob, Alice must consult the parameters made public by Bob and do the following:

- Generate a random s in **N.**
- Calculate $D = G^s$.
- Calculate $E = D^{-1}AD$.
- Calculate $K = D^{-1}BD$.

Enciphering Procedure: When the above parameters are calculated, Alice enciphers μ via

$$\mu' = K \mu K$$

and sends μ' and E to Bob.

Deciphering Procedure: When Bob receives μ' and E he does the following:

Calculates

$$L = C^{-1}EC$$

and deciphers μ' via

$$\mu = L \mu' L$$

Why the deciphering works:
The deciphering works since

$L = C^{-1}EC$
$= C^{-1}(D^{-1}AD)C$
$= D^{-1}(C^{-1}AC)D$: (D being a power of C commutes with C)
$= D^{-1}(C^{-1}A^{-1}C)^{-1}D$

$= D^{-1}B^{-1}D$: (recall that $B = C^{-1}A^{-1}C$)
$= (D^{-1}BD)^{-1}$
$= K^{-1}$: (Bob's enciphering key)

so that

$$\begin{aligned} L\,\mu'\,L &= L(K\,\mu\,K)L \\ &= (K^{-1}K)\,\mu\,(KK^{-1}) \\ &= \mu \end{aligned}$$

Wherein lies the security of the Cayley-Purser algorithm?

To find the secret matrix **C**, known to Bob alone, one might attempt to solve either the equation

$$B = C^{-1}A^{-1}C$$

or

$$G = C^{r}$$

In the first of these equations the matrix **B** is public and the matrix $\mathbf{A}^{-1}$ can be computed since both the matrix **A** and the modulus n are public.

In the second equation only the matrix **G** is known and it is required to solve for both the exponent r and the base matrix **C.** Assuming that one knew r, solving this equation would involve extracting the rth-roots of a matrix modulo the composite integer n. Even in the simplest case, where $r = 2$, extracting the square root of a 2×2 matrix modulo n requires that one be able to solve the ordinary *quadratic* congruence

$$x^2 = a \pmod{n}$$

when $n = pq$. It is known that the ability to solve this "square root" problem is equivalent to being able to factor n. Thus we may regard an attack on C via the public parameter G as being computationally prohibitive.

Solving the equation

$$B = C^{-1}A^{-1}C$$

would appear the easier option for an attack on the private matrix **C** as it involves only solving the set of linear equations given by

$$CB = A^{-1}C$$

However, the number of possible solutions to this equation is given by the order of *Cent*(A), the centralizer of A in GL(2, $\mathbf{Z}_n$). By ensuring that the order of this group is extremely large one can make it computationally prohibitive to search for C.

To see why this is the case suppose that

$$B = C^{-1}A^{-1}C \textit{ and } B = C_1^{-1}A^{-1}C_1$$

Then

$$C^{-1}A^{-1}C = C_1^{-1}A^{-1}C_1$$

if and only if

$$A^{-1}CC_1^{-1} = CC_1^{-1}A^{-1}$$

if and only if

$$CC_1^{-1} \text{ in } Cent(A^{-1})$$

if and only if

$$C \text{ in } Cent(A^{-1})C_1$$

Thus the number of distinct solutions of the equation is given by |Cent(A)| as $\text{Cent}(A^{-1}) = \text{Cent}(A)$.

Now Cent(A) will have a large order if the matrix element A has a large order.

By choosing our primes p and q to be of the form

$$p = 2p^1 + 1$$

and

$$q = 2q^1 + 1$$

where p^1 and q^1 are themselves prime, we can show that it is almost certainly the case that an element chosen at random from $GL(2, \mathbf{Z}_n)$ has a large order.

To see why, we begin by considering the homomorphism f of $GL(2, \mathbf{Z}_n)$ onto $\mathbf{Z}_n$ defined by sending a matrix into its determinant. The order of a matrix in $GL(2, \mathbf{Z}_n)$ is at least that of the order of its image in $\mathbf{Z}_n$ since . . .

If r is the order of A in $GL(2, \mathbf{Z}_n)$ and $f(A) = u$ then $A^r = I$ with

$$1 = f(I) = f(A^r) = f(A)^r = u^r$$

shows that m divides r where m is the order of u in $\mathbf{Z}_n$.

Thus the order of A in $GL(2, \mathbf{Z}_n)$ is at least m. In fact

$$f(A^m) = f(A)^m = u^m = 1$$

shows that A^m lies in $SL(2, \mathbf{Z}_n)$ so the matrix **A** will have order m iff $A^m = I$ in $SL(2, \mathbf{Z}_n)$.

We note also that since the maximum achievable order of an element in $\mathbf{Z}_n$ is

$$[p - 1, q - 1] \leq (p - 1)(q - 1) / 2 = \phi(n) / 2$$

(as $(p - 1)(q - 1) \geq 2$) and since the order of $SL(2, \mathbf{Z}_n)$ is $n\phi(n)$

$(p + 1)(q + 1)$ the maximum achievable order of a matrix in $GL(2, \mathbf{Z}_n)$ is

$$[p - 1, q - 1]n\phi(n)(p + 1)(q + 1) \Rightarrow$$
$$n\phi(n)^2(p + 1)(q + 1) / 2 = |GL(2, \mathbf{Z}_n)| / 2$$

Thus if we can show that the probability of an element having a small order in $\mathbf{Z}_n$ is negligibly small then we will have shown that the order of an element chosen at random from $GL(2, \mathbf{Z}_n)$ is almost certainly of "high order."

If

$$p = 2p^1 + 1$$

and

$$q = 2q^1 + 1$$

then

$$\phi(n) = \phi(pq) = (p - 1)(q - 1) = 2p^1 2q^1 = 4p^1 q^1$$

with

$$[p - 1, q - 1] = [2p^1, 2q^1] = 2p^1 q^1 = \phi(n)/2$$

Now the possible orders of the elements in $\mathbf{Z}_n$ are divisors of $\phi(n)/2 = 2p^1q^1$ and so are

$$1, 2, p^1, q^1, 2p^1, 2q^1, p^1q^1, 2p^1q^1$$

and all of these orders are achieved by some elements. In fact, by counting exactly how many elements correspond to each order we show that the probability of finding a unit in $\mathbf{Z}_n$ *of order less than* p^1q^1 is negligibly small.

Recall that if a in $\mathbf{Z}_p$ has order k and b in $\mathbf{Z}_q$ has order l then the order of c in $\mathbf{Z}_n$ where

$$c \equiv a \pmod{p}$$

and

$$c \equiv b \pmod{q}$$

is $[k, l]$, the least common multiple of k and l.

Now the possible orders of a and b in $\mathbf{Z}_p$ and $\mathbf{Z}_q$ are divisors of

$$p - 1 = 2p^1 \quad ; \quad q - 1 = 2q^1$$

respectively.

The following table lists the possible orders along with the number of elements of each order.

$\mathbf{Z}^*_p$		$\mathbf{Z}^*_q$	
Possible Orders	No. of elements of that order	Possible Orders	No. of elements of that order
1	1	1	1
2	1	2	1
p^1	$p^1 - 1$	q^1	$q^1 - 1$
$2p^1$	$p^1 - 1$	$2q^1$	$q^1 - 1$

By lifting elements in pairs via the CRT we obtain the elements corresponding to the different orders in $\mathbf{Z}_n$ along with number of elements of each order.

Order	Number	Reason
1	1	$[1, 1] = 1$
2	3	$[1, 2] = [2, 1] = [2, 2] = 2$
p^1	$p^1 - 1$	$[p^1, 1] = p^1$
q^1	$q^1 - 1$	$[1, q^1] = q^1$
$2p^1$	$3p^1 - 3$	$[2p^1, 1] = [p^1, 2] = [2p^1, 2] = 2p^1$
$2q^1$	$3q^1 - 3$	$[1, 2q^1] = [2, q^1] = [2, 2q^1] = 2q^1$
p^1q^1	$p^1q^1 - p^1 - q^1 + 1$	$[p^1, q^1] = p^1q^1$
$2p^1q^1$	$3p^1q^1 - 3p^1 - 3q^1 + 3$	$[2p^1, q^1] = [p^1, 2q^1] =$ $[2p^1, 2q^1] = 2p^1q^1$

Note that if we sum all the individual counts we get exactly $4p^1q^1$ which is the number of elements of $\mathbf{Z}_n$.

Explanation: To see how the number of elements corresponding to an order is obtained consider the last entry in the above array: An element of order $2p^1q^1$ in $\mathbf{Z}_n$ can be obtained in three different ways by lifting pairs of elements from $\mathbf{Z}_p$ and $\mathbf{Z}_q$: one way is lifting the pair (a, b) where a has an order $2p^1$ and b has order q^1; another by lifting the pair (a, b) where a has an order p^1 and b has order $2q^1$ and another by lifting the pair (a, b) where a has an order $2p^1$ and b has order $2q^1$.

Regarding elements of order less than p^1q^1 as elements of "low order" we obtain the probability of choosing an element of order less than p^1q^1 to be

$$\frac{4p^1 + 4q^1 - 4}{4p^1q^1}$$

This is equivalent to

$$\frac{1}{p^1} + \frac{1}{q^1} - \frac{1}{p^1 q^1}$$

In the case where p and q are both of order of magnitude 10^{100} this probability is approximately

$$2.10^{-100}$$

which, by any standards, is negligibly small.

Some differences between the RSA and Cayley-Purser Algorithms

1. The most significant difference between the RSA and Cayley-Purser algorithms is the fact that the Cayley-Purser algorithm uses only modular matrix multiplication to encipher plaintext messages whereas the RSA uses modular exponentiation which requires considerably longer computation time. Even with the powerful *Mathematica* function PowerMod the RSA appears (see Tables I–IX) to be over twenty times slower than the Cayley-Purser algorithm.
2. In the RSA the parameters needed to encipher—(n, e)—are published for the whole world to see and anyone who wishes to send a message to Bob raises his messages' numerical equivalents to the power of *e modulo n*. However, in the Cayley-Purser algorithm the enciphering key is not made public! Only the parameters for calculating one's own key are published. This means that every sender in this system also enjoys a certain measure of secrecy with regard to her own messages. One consequence of this is that the Cayley-Purser algorithm is not susceptible to a repeated encryption attack because the sender, Alice, is the only one who knows the encryption key she used to encipher. In the

RSA, however, if the order e can be found then an eavesdropper can decipher messages.

3. Alice can choose to use a new enciphering key every time she wishes to write Bob. In the unlikely event that an eavesdropper, Eve, should find an enciphering key, she gains information about only one message and no information about the secret matrix **C.** By contrast, if a piece of intercepted RSA ciphertext leads to Eve being able to decipher (through repeated encryption, etc.), then she would be able to decipher all intercepted messages which are enciphered using the public exponent e.
4. In the Cayley-Purser algorithm the sender, Alice, has the ability to decipher the ciphertext which she generates using Bob's public parameters even if she loses the original message (because she knows D and therefore can get the deciphering key, $K^{-1} = L$!). Contrast this to the RSA—Alice cannot decipher her own message once she has enciphered it using Bob's public key parameters. There is a possible advantage in this for Alice in that she could store encrypted messages on her computer ready for sending to Bob.

RSA vs. Cayley-Purser

Empirical Time Analysis

The times taken by the Cayley-Purser and RSA algorithms (using a modulus n of the order 10^{200}) to encipher single and multiple copies of the *Desiderata* (1769 characters) by Max Ehrman are given in the following tables along with the times taken by both algorithms to decipher the corresponding ciphertext.

Table I

Running Time (Seconds) Message = 1769 characters				
Trial No.	1	2	3	Average
RSA encipher	41.94	42.1	41.78	41.94
RSA decipher	40.99	41.009	41.019	41.009
C-P encipher	1.893	1.872	1.893	1.886
C-P decipher	1.502	1.492	1.492	1.4953

Table II

Running Time (Seconds) Message = 2 * 1769 = 3538 characters				
Trial No.	1	2	3	Average
RSA encipher	72.364	72.274	72.364	72.334
RSA decipher	70.942	70.952	72.144	71.346
C-P encipher	3.305	3.305	3.325	3.3016
C-P decipher	2.734	2.864	2.864	2.8206

Table III

Running Time (Seconds) Message = 3 * 1769 = 5307 characters				
Trial No.	1	2	3	Average
RSA encipher	103.078	102.808	103.489	103.125
RSA decipher	101.246	101.076	104.06	102.1273
C-P encipher	4.757	4.737	4.747	4.747
C-P decipher	3.976	4.086	4.066	4.0426

Table IV

Running Time (Seconds) Message = 4 * 1769 = 7076 characters				
Trial No.	1	2	3	Average
RSA encipher	134.434	134.323	134.333	134.363
RSA decipher	131.128	134.734	134.734	133.532
C-P encipher	6.159	6.048	6.109	6.1053
C-P decipher	5.227	4.967	4.967	5.05536

Table V

Running Time (Seconds) Message = 12 * 1769 = 21,228 characters				
	RSA enc	RSA dec	C-P enc	C-P enc
Time Taken	378.078	371.254	17.435	14.371

Table VI

Running Time (Seconds) Message = 24 * 1769 = 42,456 characters				
	RSA enc	RSA dec	C-P enc	C-P enc
Time Taken	509.523	511.455	22.583	18.767

Table VII

Running Time (Seconds) Message = 48 * 1769 = 84,912 characters				
	RSA enc	RSA dec	C-P enc	C-P enc
Time Taken	1019.24	1023.95	44.894	36.823

Table VIII

Running Time (Seconds) Message = 144 * 1769 = 254,736 characters				
	RSA enc	RSA dec	C-P enc	C-P enc
Time Taken	3154.21	3036.24	142.775	129.416

With respect to a 133MHz machine the Cayley-Purser algorithm is on average approximately twenty-two times faster than the RSA where in each case the modulus n is of the order 10^{200}.

Table IX

The following table illustrates the time taken for the RSA and CP algorithms to encipher a piece of text (7076 characters in length) with varying size moduli. The ratio of the enciphering speeds is also given.

Running Time (Seconds) Message μ containing 7076 characters			
Modulus	RSA	CP	Ratio
222 digits	84.641	3.916	21.6:1
242 digits	104.71	4.036	25.9:1
262 digits	118.841	4.276	27.8:1
282 digits	131.739	4.326	30.5:1
302 digits	145.689	4.487	32.5:1

Note: the difference in times taken to encipher and decipher in the RSA depends on the binary weight of the exponents *e* and *d*.

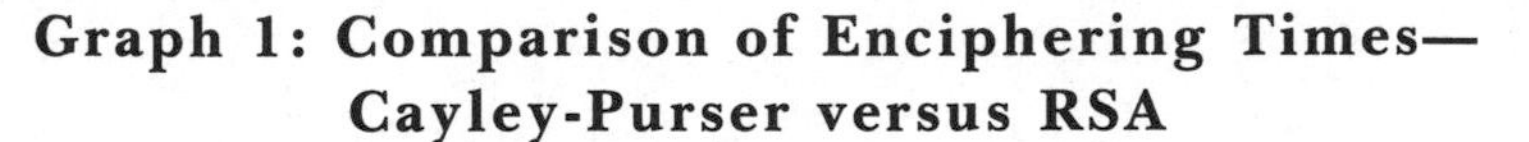

Graph 1: Comparison of Enciphering Times—Cayley-Purser versus RSA

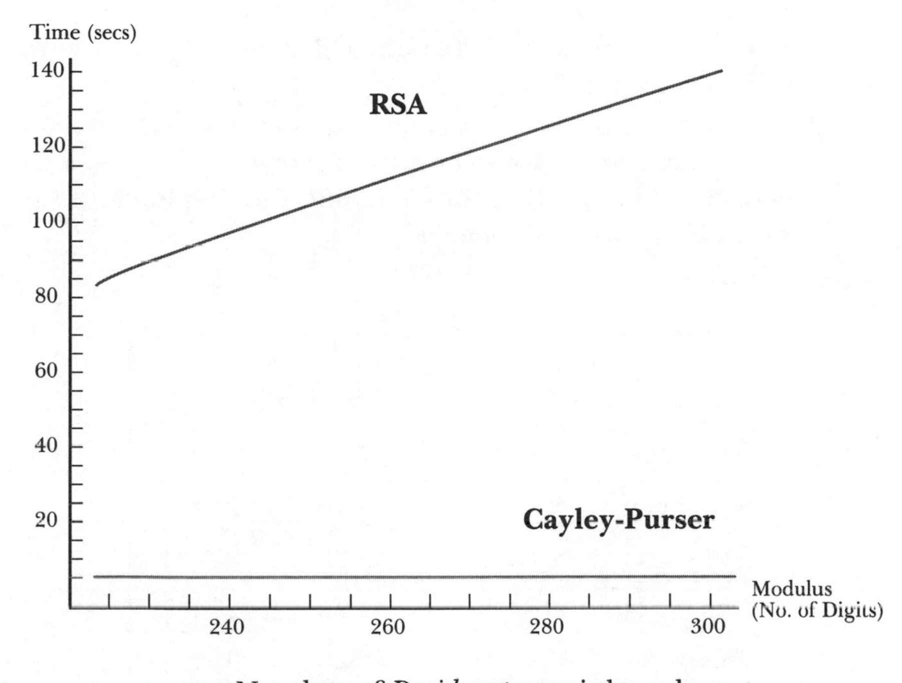

Number of *Desiderata* enciphered
The piece of text used (*Desiderata*) contains 1769 characters.

Conclusions

This project

- (a) shows mathematically that the CP algorithm is as secure as the RSA algorithm.
- (b) illustrates through an empirical run-time analysis that the CP algorithm is FASTER to implement than the RSA algorithm: the speed factor increasing with modulus size as shown in the following table:

Running Time (Seconds) Message = 4 * 1769 = 7076 characters			
Modulus	RSA	CP	Ratio
222 digits	84.641	3.916	21.6:1
242 digits	104.71	4.036	25.9:1
262 digits	118.841	4.276	27.8:1
282 digits	131.739	4.326	30.5:1
302 digits	145.689	4.487	32.5:1

Postscript: An Attack on the CP Algorithm

We describe an attack on the Cayley-Purser algorithm which shows that anyone with a knowledge of the public parameters A, B and G can form a multiple C' of C. This matrix **C'** can then be used in conjunction with E to form $L = K^{-1}$ which is the deciphering key. Thus the system as originally set out is "broken."

If $C' = vC$ for some constant v and if E is known to an adversary then the calculation

$$C'^{-1}EC = (v^{-1}C^{-1})\,E\,(vC) = C^{-1}EC = K^{-1}$$

yields the deciphering key K^{-1}. Thus any multiple of C can be used to decipher.

In the CP system the matrix G is made to commute with C so as to enable the deciphering process. This is done using the construction $G = C^r$ for some r and herein lies the weakness of the algorithm. Were G to be generated more efficiently using a linear combination of C and the identity matrix I (higher order polynomials in C reduce via the Cayley-Hamilton theorem to linear expressions in C) the system is still compromised.

If the matrix G is nonderogatory (i.e., when G is reduced mod p and mod q, neither of the two matrices obtained is a scalar multiple of the identity) then

$$C = uI + vG$$

(If the matrix G is derogatory then n can be factored by calculating $GCD(g_{11} - g_{22}, g_{12}, g_{21}, n)$.)

Now since G is nonderogatory $(v, n) = 1$ and

$$C' = v^{-1}C = v^{-1}uI + G = DI + G$$

for some d in $\mathbf{Z}_n$.

Since

$$\begin{aligned} B &= C^{-1}A^{-1}C \\ &= vC^{-1}A^{-1}v^{-1}C \\ &= (v^{-1}C)^{-1}A^{-1}(v^{-1}C) \\ \Rightarrow B &= C'^{-1}A^{-1}C' \\ \Rightarrow C'B &= A^{-1}C' \end{aligned}$$

Substituting $dI + G$ for C' in this last equation gives

$$\begin{aligned} [dI + G]B &= A^{-1}[dI + G] \\ \Rightarrow d\,B + GB &= dA^{-1} + A^{-1}G \\ \Rightarrow d[B - A^{-1}] &= [A^{-1}G - GB] \end{aligned}$$

Since $A \neq B^{-1}$ these matrices differ in at least one position. For argument's sake let $a_{11} \neq b^{-1}{}_{11}$. Comparing the (1, 1) entries in the above matrix identity gives

$$d(b_{11}{}^{-1} - a_{11}) \equiv e \pmod{n} \text{ ; } e \text{ in } \mathbf{Z}_n$$

If $(a_{11} - b^{-1}{}_{11})^{-1}$ exists mod n the above linear congruence is uniquely solvable for d. If not a factorization of n is obtained.

Remark 1: This attack shows that anyone with a knowledge of the public parameters A, B and G can form a multiple C' of C. This matrix C' can then be used to form $L = K^{-1}$ *provided E is known.* If E is transmitted securely on a once off basis then knowledge of a C' on its own is not enough to break the system, though then the Cayley-Purser algorithm would no longer be public key in nature.

Remark 2: The fact that a derogatory G leads to a factorization of the modulus n was further investigated on the assumption that knowledge of n might not severely compromise the system. However in this case also a multiple of C is obtainable.

Remark 3: An analysis of the CP algorithm based on 3×3 matrices, though slightly more involved in its details, leads to conclusions similar to the ones just described.

Remark 4: For the sake of efficiency D should be calculated as $\mathbf{D} = \mathbf{aG} + \mathbf{b}I$ rather than as $D = G^s$.

Mathematica Code for RSA & CP Algorithms

```
FirstPrimeAbove[n Integer]
(Clear[k];k = n;While[! PrimeQ[k],k = k + 1];k)
```

```
ConvertString[str_String] :=
Fold[Plus[256 #1, #2]&, 0, ToCharacterCode[str]]
```

```
StringToList[text_string] := Module[{blockLength =
Floor[N[Log[256, n]]], strLength = StringLength[text]},
ConvertString /@ Table[StringTake[text,{i, Min[strLength, 1 +
blockLength – 1]}],{i, 1, strLength, blockLength}]]
```

```
ConvertNumber[num_Integer]:=
FromCharacterCode /@ IntegerDigits[num,256]
```

```
ListToString[l_List] := StringJoin[ConvertNumber /@ 1]
```

Mathematica Code for RSA Algorithm

```
GeneratePQNED[digits_Integer] := (p = FirstPrimeAbove[
  prep = Random[Integer, {10^(digits–1), 10^digits–1}]];
Catch[Do[preq = Random[Integer, 10^(digits–1),
  10^digits–1}];
If[preq[=[re[,Throw[q = FirtPrimeAbove[preq]]], {100}]]
  n = pq;e = Random[Integer, {p, n}];
While[GCD[e, (p–1) (q–1) i = 1,
  e = Random[Integer, {p, n}]]; e;
  d = PowerMod[e, –1 (p–1) (q–1)];)
```

```
RSAencNumber[num_Integer] := PowerMod[num, e, n]
```

```
RSAdecNumber[num_Integer] := PowerMod[num, d, n]
```

```
RSAenc[text_String]:=
RSAencNumber[#]& /@ StringToList[text]
```

```
RSAdec[cipher_List]:ListToString[RSAdecNumber
  [#]&/@cipher]
```

Mathematica Code for Cayley-Purser Algorithm

```
StringToMatrices[text_String]:= Partition[Parition[Flatten
[Append[StringToList[text],{32,32,32}]],2],2]
```

```
MatriceToString[l_List] :=
StringJoin [ ConvertNumber/@Flatten[1]]
```

```
CPpqn[digits_Integer] :=Module[{
  p1 = FirstPrimeAbove[Random[Integer,
    {10^(Floor[digits/2]−1), 10^(Floor[digits/2])−1}]],
  q1 = FirstPrimeAbove[Random[Integer,
    {10^(Floor[digits/2]−1), 10^(Floor[digits/2])−1}]],
While[PrimeQ[p = 2p1 +1], p1 =
  FirstPrimeAbove[p1 + 1]]; p;
While[PrimeQ[q = 2q1 +1], q1 =
  FirstPrimeAbove[q1 + 1]];
q; n = pq; ]
```

```
randmatrix := (Catch[
  Do[m = Table[Random[Integer, {0, n}], {i, 1, 2},
  {j, 1, 2}];
If[GCD[Mod[Det[m], n], n] == 1, Throw[m]], {1000}]])
```

```
inv[a_] := (d = Mod[Det[a], n]; i = PowerMod[d, −1, n];
{{Mod[i * a[[2, 2]], n], Mod[−i * a[[1, 2]], n]},
  {Mod[−i * a[[2, 1]], n], Mod[i * a[[1, 1]], n]}})
```

```
mmul[j_, k_] := Mod[
  {{Mod[j[[1, 1]]*k[[1, 1]], n] + Mod[j[[1, 2]]*k[[2, 1]], n],
    Mod[j[[1, 1]]*k[[1, 2]], n] + Mod[j[[1, 2]]*k[[2, 2]], n]},

    {Mod[j[[2, 1]]*k[[1, 1]], n] + Mod[j[[2, 2]]*k[[2, 1]], n],
      Mod[j[[2, 1]]*k[[1, 2]], n] + Mod[j[[2, 2]]*k[[2, 2]],
      n]}}, n]
```

```
CPparameters := (identity = {{1, 0}, {0, 1}};
  alpha = randmatrix; Catch[Do[chi = randmatrix;
If[mmul[chi, alpha] ! = mmul[alpha, chi],
    Throw[chi]], {10000000}]]
  chiinv = inv[chi]; alphainv = inv[alpha];
  Catch[Do[s = Random[Integer, {2, 50}];
    gamma =Mod[MatrixPower[chi, s], n];
  If[gamma != identity, Throw[gamma]], {10000000}]];
Catch[Do[delta = Mod[Mod[Random[Integer, {1,
  n−1}]gamma, n]
    + Mod[Random[Integer, {1, n−1}]identity, n], n]
    If[delta !=identity &&
      mmul[delta, alpha] !=mmul[alpha, delta], Throw
      [delta]],
    {10000000}];
  beta = mmul[mmul[chiinv, aphainv], chi];
  deltainv = inv[delta];
  epsilon = mmul[mmul[deltainv, alpha], delta];
kappa = mmul[mmul[deltainv, beta], delta];
  lamda = mmul[mmul[chiinv, epsilon], chi];)
```

```
CPenc[plain_String] := CPencNum [ StringToMatrices[plain]]
```

```
CPDecNum[l_list] := Table[mmul[mmul[lamda, l[[i]]],
  lamda], {i, Length[l]}]
```

```
CPEncNum[l_List] :=
  Table[mmul[mmul[kappa, l[[i]]], kappa], {i, Length[l]}]
```

```
CPdec[cipher_List] := MatricesToString[CPDecNum[cipher]]
```

Bibliography:

Higgins, J. and Cambell, D.: Mathematical Certificates. Math. Mag 67 (1994). 21-28.

Mackiw, George: Finite Groups of 2 × 2 Integer Matrices. Math. Mag 69 (1996). 356-361.

Meijer, A.R.: Groups, Factoring and Cryptography. Math. Mag 69 (1996). 103-109.

Menezes, van Oorschot, Vanstone: Handbook of Applied Cryptography, CRC Press 1996.

Salomaa, Arto: Public-Key Cryptography (2 ed.). Springer Verlag 1996.

Schneier, Bruce: Applied Cryptography. Wiley 1996.

Stangl, Walter D.: Counting Squares in Z_n. Math. Mag 69 (1996). 285-289.

Sullivan, Donald: Square Roots of 2 × 2 Matrices. Math. Mag 66 (1993). 314-316.

Appendix B

Answers to Miscellaneous Questions

2 Early Challenges

Page 25: *If so, the cell marked with a * must contain the number 1. Why?*

Because 9, 5 and * must add up to 15.

Page 26: *The numbers* A *and* B *in the first column cannot be any of 6, 7 or 8. Why?*

Because each, when added to the 9 in the central side cell, gives a sum of at least 15 with just two numbers.

5 Of Prime Importance

Page 48: *Can you find the number less than 1000 which has the most factors?*

I suppose you can answer this question only with a yes or a no.

However, the number is 840, which has 32 factors. The number 720 has 30 factors, while the number 960 has 28 factors.

Page 50: *Are there other rectangular arrangements (of the number 889,364)?*

Yes. Here are all the possible rectangular arrangements, beginning with the trivial one:

1	×	889,364
2	×	444,682
4	×	222,341
7	×	127,052
14	×	63,526
23	×	38,668
28	×	31,763
46	×	19,334
92	×	9667
161	×	5524
322	×	2762
644	×	1381

Page 59: *Do you now see why the first 120 numbers were chosen to demonstrate this remarkably simple algorithm?*

Again, only you can answer this question with a yes or a no. However, the first 120 numbers were chosen because sievings based on the four primes 2, 3, 5 and 7 sieve out all the composites less than or equal to 120, but do not sieve out the composite 121. This requires a sieving based on the fifth prime, 11, because 121 = 11 × 11.

Page 60: *Prove that there are at least two people in Dublin with the same number of hairs on their head.*

Try to imagine a bighead, and make a rough estimate of the number of hairs that could be on that head. I suppose 2 per mm^2 is a bit light; let's overestimate it at an enormous 10 per mm^2, which is 1000 per cm^2. This is pretty hairy.

Now for the size of this head: suppose it's a sphere with about half its surface area covered with hair. The area of a hemisphere of radius R is $2\pi R^2$. What would R be for a bighead? Well, my basketball has a diameter of 23.25 cm. Let's suppose a bighead's diameter to be a whopping 30 cm, so its radius R would be 15 cm. So the hairy bighead area is

$$2 \times \pi \times 15^2 \approx 1414$$

rounding up to 1450 cm^2. How many hairs sprout from 1450 cm^2 if there are 1000 per cm^2? Answer:

$$1450 \times 1000 = 1{,}450{,}000$$

I hope you will agree that this is an absolute maximum.

Now, how many people live in Dublin? I'm told at least 1,500,000. Let's settle for that. Suppose now that a survey is done on the number of hairs on the head of each Dubliner. Suppose also that (by some miracle) of the first 1,450,001 people surveyed each has a different number of hairs. Now all the possible hair numbers which lie between 0 and 1,450,000 have been used. Any one of the remaining 49,999 citizens also has a hair number which lies between 0 and 1,450,000 and so matches the hair number of one of the persons in the first 1,450,001 people surveyed. So there must be at least two people with the same number of hairs on their head.

Page 66: *Check that* M(10) $= 2^{10} - 1 = 1023$ *is indeed composite.*

The digits of the number 1023 add to 6. Since this is divisible by 3, then so is the number 1023. Dividing 1023 by 3 gives 341. Since $341 = 11 \times 31$, the complete factorization of 1023 is $3 \times 11 \times 31$.

Page 67: *$2^{11} - 1 = 2047$ is not a prime.*

Its smallest prime factor is 23 and its remaining prime factor is 89. Observe that $23 = 2 \times 11 + 1$ and $89 = 8 \times 11 + 1$. Curious?

6 The Arithmetic of Cryptography

Page 80: *Decipher* **wubdjdlq** *assuming it has been enciphered using a 3-shift.*

Answer: **tryagain**, taken as "try again."

Occasionally a word becomes another recognizable English word when it is enciphered.

The word "cold" enciphers to "frog" under a 3-shift.

What happens to "pecan" in a 4-shift cipher?

The word "pecan" enciphers to "tiger" under a 4-shift.

What happens to "sleep" in a 9-shift cipher?

The word "sleep" enciphers to "bunny" under a 9-shift.

Page 94: *Things are much worse when every* P-*number is multiplied by 13 modulo 26. Do a few multiplications and you'll see why.*

The following array shows the results of enciphering the numbers 0 to 25 according to the rule $C = 13P \bmod 26$:

P	$\rightarrow$	C	P	$\rightarrow$	C
0	$\rightarrow$	0	13	$\rightarrow$	13
1	$\rightarrow$	13	14	$\rightarrow$	0
2	$\rightarrow$	0	15	$\rightarrow$	13
3	$\rightarrow$	13	16	$\rightarrow$	0
4	$\rightarrow$	0	17	$\rightarrow$	13
5	$\rightarrow$	13	18	$\rightarrow$	0
6	$\rightarrow$	0	19	$\rightarrow$	13
7	$\rightarrow$	13	20	$\rightarrow$	0
8	$\rightarrow$	0	21	$\rightarrow$	13
9	$\rightarrow$	13	22	$\rightarrow$	0
10	$\rightarrow$	0	23	$\rightarrow$	13
11	$\rightarrow$	13	24	$\rightarrow$	0
12	$\rightarrow$	0	25	$\rightarrow$	13

When you have found the inverse of 7 modulo 26, try to find the inverse of 4 modulo 26.

The inverse of 7 modulo 26 is 15, because $15 \times 7 = 105$ and 105 mod $26 = 1$. The number 4 has no inverse modulo 26. There is no whole number between 0 and 25 which, when multiplied by 4 and reduced modulo 26, yields 1. Just look at the array showing the results of enciphering the numbers 0 to 25 according to the rule $C = 4P \bmod 26$.

Page 99: *Then the* P-*number assigned to* **go** *is*

$$P = (6 \times 26) + 14 = 170$$

Can you explain why this is correct?

Once again, only you can answer this question with a yes or a no. However, the group of digraphs with leading character **g**, beginning **ga, gb, gc** and ending **gx, gy, gz,** is the *seventh* group, because *g* is the seventh character of the lower-case alphabet. The first six groups contain $6 \times 26 = 156$ digraphs, which are labeled with the numbers 0 to 155. Since *o* is the fifteenth character of the lower-case alphabet, the digraph **go** is the fifteenth in its group and so has the number 155 + 15 = 170.

7 Sums with a Difference

Page 117: *If the number 675 represents a Wednesday, which day of the week does the number 943 represent?*

Since 943 − 675 = 268 = 38 × 7 + 2, the number 943 represents a Friday because it is 38 weeks and 2 days after the Wednesday represented by the number 675. The alert reader will have noticed that this question is posed in a slightly different form—and answered—a few paragraphs previously.

Page 127: *Then add this 8 to 6 to get 5. Why?*

Because 8 + 6 = 14 and 14 mod 9 = 5, as 14 = 1 × 9 + 5.

Page 139: *Can you find two nontrivial factors of 11,111?*

Again, over to you for a yes or a no. However, 41 × 271 is the complete factorization of 11,111.

Page 148: *Should you switch?*

Yes. In order to think clearly about the problem get someone to help you simulate the game show with three mugs (to act as the doors) and a matchstick (to act as the car). Close your eyes and get your helper to hide the matchstick under one of the mugs at random. Then open your eyes, choose a mug and let your helper reveal a mug with nothing under it. Play this game many times, and count how often you win by not switching and how often by switching.

The following explanation of the answer "Yes, you should

switch" is from Erich Neuwirth of Bad Voeslau, Austria, and appears on page 369 of the Mathematical Association of America's *The College Mathematics Journal,* vol. 30, no. 5, November 1999:

> Imagine two players, the first one always staying with the selected door and the second one always switching. Then, in each game, exactly one of them wins. Since the winning probability for the strategy "Don't switch" is ⅓, the winning probability for the second one is ⅔ and therefore switching is the way to go.

8 One Way Only

Page 152: *Can you find* p *and* q*?*

No!

Do you want to?

Probably not.

Page 156: *What could you say about the oddness or otherwise of* x *if you were to receive a* y *that is even?*

That x is even because (in ordinary arithmetic) an odd number cubed is always an odd number.

9 Public Key Cryptography

Page 165: *For 10 people, 45 such meetings are needed. Why?*

Imagine each of the 10 people shaking hands. Each person would shake the hand of each of the other 9. Since there are 2 people involved in each handshake, the total number of handshakes is 45 as this number is one-half of $10 \times 9 = 90$.

Page 183: *You might like to help Alice, Bob and Claire figure out what to do if they wish Denis to join their team and still maintain a threshold of two, meaning that any two of the team of four can work without the others being present. In other words, they need you to help them come up with a (2,4) scheme. Will they have to buy more locks, or can they get away with making more keys to the locks they already have?*

Yes and no. They will have to buy one more lock. They cannot get away with just making more keys to the three locks they already have. The reason is simple. Because they require a (2,4) scheme it must be that each one of them is unable to open (at least) one lock. If there are only three locks, then since there are four people some two of them, say Alice and Bob, cannot open the *same* lock, L_1 (lock one) for example. However, this means that there is one group of two people unable to work, namely Alice and Bob, whereas the intention is that any group of two can open the cabinet to begin work. Hence at least four locks are needed with each member of the group unable to open a lock different from the ones the other members are unable to open. In fact, four locks will do if each member is given three keys according to the following scheme (the four locks are labeled L_1, L_2, L_3 and L_4 respectively):

> Alice has a key for each of L_1, L_2 and L_3 but not for L_4.
> Bob has a key for each of L_1, L_2 and L_4 but not for L_3.
> Claire has a key for each of L_1, L_3 and L_4 but not for L_2.
> Denis has a key for each of L_2, L_3 and L_4 but not for L_1.

It is clear that for any two people, each has the key the other is missing.

Appendix C

Euclid's Algorithm

The greatest common divisor (g.c.d.) of any two natural numbers is the largest natural number that divides each of them.

In general, the notation (c, d) stands for, and is read as, the greatest common divisor of the numbers c and d. For example, $(8, 12) = 4$ says that the greatest common divisor of 8 and 12 is 4, while $(30, 78) = 6$ says that the greatest common divisor of 30 and 78 is 6. On the other hand, $(10, 21) = 1$ because 10 and 21 have no factor in common other than 1. The numbers 10 and 21 are *relatively prime* to each other.

Surprisingly, $(m, 0) = m$ for any natural number m. Why? Because every m is a factor of 0 since $0 = 0 \times m$. Thus, for example, $(30, 0) = 30$, while $(11, 0) = 11$.

But a nice notation is not enough to tell us how to find the g.c.d. of two numbers. For example, what is (4950, 420)? To put it in words, what is the g.c.d. of the two numbers 4950 and 420?

Because both numbers end in 0, you might say that—at the very least—they have the factor 10 in common. Having made this observation (which won't always be valid), you might examine the numbers 495 and 42 to see which factors they have in common. After a little thought, it might occur to you to find the complete factorization of both of the original numbers, pick out those factors in common and form their product to get the desired greatest common divisor.

Since

$$4950 = 2 \times 3^2 \times 5^2 \times 11$$
$$420 = 2^2 \times 3 \times 5 \times 7$$

the two numbers share the factors 2, 3 and 5. Since $2 \times 3 \times 5 = 30$, it must be that

$$(4950, 420) = 30$$

That wasn't too difficult. But fundamentally, as an approach, it is very seriously flawed because it requires the complete factorization of the two numbers whose g.c.d. is sought. Since factorization is believed, with some justification, to be a tremendously difficult problem in general, this method cannot be considered a useful one. To convince yourself on this point, write down two random 100-digit numbers and try to find their g.c.d.

Euclid's Algorithm

Now for a bit of good news. Over two thousand years ago Euclid described a very efficient method for finding the g.c.d. of any two natural numbers, which *does not require* knowledge of their factorizations. The method is based on a simple general observation which can be illustrated using the two "smallish" numbers 4950 and 420. The observation is

$$(4950, 420) = (420, 4950 \bmod 420)$$

Let us unpack this a little to see what is being said. Since

$$4950 = 11 \times 420 + 330 \Rightarrow 4950 \bmod 420 = 330$$

the observation simplifies to

$$(4950, 420) = (420, 330)$$

What is this all-important fact, which I'm asking you to take on trust as being true, saying? It says that the g.c.d. of the original pair of numbers {4950, 420} is *exactly the same* as the g.c.d. of the pair of numbers {420, 330}. This is a very big deal. Why? Because the second pair of numbers is much friendlier than the first pair—the new number 330 in the second pair is smaller than the first number 4950 of the original pair. It should be easier to find (420, 330) rather than (4950, 420).

Now for a brilliant stroke. The original observation applied to (420, 330) says that

$$(420, 330) = (330, 420 \bmod 330)$$

or that

$$(420, 330) = (330, 90)$$

since $420 = 1 \times 330 + 90$ gives $420 \bmod 330 = 90$. Hence

$$(4950, 420) = (420, 330) = (330, 90)$$

Common sense suggests that we do exactly to (330, 90) what has already been done to (4950, 420) and (420, 330). Another such "iteration" gives

$$(4950, 420) = (420, 330) = (330, 90) = (90, 60)$$

since $330 = 3 \times 90 + 60 \Rightarrow 330 \bmod 90 = 60$. It is possible to stop now, since $(90, 60) = 30$ by inspection, and conclude that $(4950, 420) = 30$. This is the same answer as we obtained previously.

Streamlining all the steps required in the above procedure (and doing two more) gives

$$\begin{aligned}
4950 &= 11 \times 420 + 330 \\
420 &= 1 \times 330 + 90 \\
330 &= 3 \times 90 + 60 \\
90 &= 1 \times 60 + 30 \\
60 &= 2 \times 30 + 0
\end{aligned}$$

from which the series of equalities

$$(4950, 420) = (420, 330) = (330, 90) = (90, 60) = (60, 30) = (30, 0)$$

follow. Since $(30, 0) = 30$, the g.c.d. of the numbers 4950 and 420 is 30.

The reason for carrying the calculations as far as we have is that a program implementing the procedure stops when it encounters the remainder 0 in the last line. The last nonzero remainder appearing in the previous line is the desired g.c.d. of the two numbers.

This magnificently simple and efficient procedure, which uses nothing but basic arithmetic, has been known for centuries as *Euclid's algorithm,* or the *Euclidean algorithm.* With very slight modifications it is still the algorithm used by today's "number crunchers" to find the greatest common divisor of two numbers. This algorithm always terminates because the number of possible remainders is at most the size of the smaller of the two numbers whose g.c.d. is sought.

Euclid's algorithm can be used to find the multiplicative inverse of a number with respect to a modulus (whenever the number possesses such an inverse). The fact that the algorithm does this very efficiently is of some importance. In Chapter 6, "The Arithmetic of Cryptography," the multiplicative inverse of 5 modulo 26 was needed in order to obtain the deciphering rule corresponding to the enciphering rule $C = 5P \bmod 26$. This inverse is obtained by first using Euclid's algorithm to determine (26, 5), even though it is obvious without calculation that the answer is 1. Imitating the procedure discussed above, the following array of calculations is obtained:

$$26 = 5 \times 5 + 1$$
$$5 = 5 \times 1 + 0$$

Since the last nonvanishing remainder is 1, it's the case that (26, 5) = 1. Having determined that (26, 5) = 1, we begin a second procedure of writing the g.c.d. 1 in terms of 26 and 5. When this is done, we'll explain why it is a useful thing to do. The first line of the array

$$26 = 5 \times 5 + 1$$

can be rearranged to give

$$1 \times 26 + (-5) \times 5 = 1$$

This last equation shows that 1 can be written as a combination of 26 and 5. Specifically, it shows that 1 can be made by combining one 26 with minus five 5's. Writing the equation as

$$(-5) \times 5 = -1 \times 26 + 1$$

gives

$$(-5 \times 5) \bmod 26 = 1$$

This is very significant. It says that when the number 5 is multiplied by −5 and the result reduced modulo 26, the number 1 is obtained. So −5 is a multiplicative inverse of 5 modulo 26. Because −5 is not in the range 0 to 25, it is not regarded as the multiplicative inverse of 5 modulo 26. However, since −5 mod 26 = 21, *the* multiplicative inverse of 5 modulo 26 is taken to be 21. It is easy to confirm that 21 is a multiplicative inverse of 5 modulo 26, as follows:

$$21 \times 5 = 105 = 4 \times 26 + 1 \Rightarrow (21 \times 5) \bmod 26 = 1$$

Thus has Euclid's algorithm been used to find the multiplicative inverse of 5 modulo 26. Hence the deciphering rule corresponding to the enciphering rule $C = 5P \bmod 26$ is

$$P = 21C \bmod 26$$

Knowing that 21 is the multiplicative inverse of 5 allows us (by glossing over some details) to sketch an explanation of why

$$P = (21C + 12) \bmod 26$$

is the deciphering rule corresponding to the enciphering rule

$$C = (5P + 18) \bmod 26$$

Adding -18 to both sides of this equation gives

$$(C - 18) = 5P \bmod 26$$

or

$$5P = (C + 8) \bmod 26$$

since $-18 \bmod 26 = 8$. Multiplying both sides of this equation by 21, which is the multiplicative inverse of 5, and reducing 105 modulo 26 to 1 gives

$$P = 21\,(C + 8) \bmod 26$$

Since $21 \times 8 = 168 = 6 \times 26 + 12$, the above can be written as

$$P = (21C + 12) \bmod 26$$

This, then, is the deciphering rule corresponding to the enciphering rule

$$C = (5P + 18\,) \bmod 26$$

Finding Multiplicative Inverses

The calculations involved in finding the multiplicative inverse of 5 modulo 26 are a little too short to reveal the full extent of what must often be done to determine the multiplicative inverse of a number. Greater insight into the method is gained by using it to compute the multiplicative inverse of 7 modulo 26. The following array of calculations:

$$\begin{aligned} 26 &= 3 \times 7 + 5 \\ 7 &= 1 \times 5 + 2 \\ 5 &= 2 \times 2 + 1 \\ 2 &= 2 \times 1 + 0 \end{aligned}$$

shows that $(26, 7) = 1$.

We now begin the second stage of the procedure for finding the multiplicative inverse of 7 modulo 26. This requires that we "work back up the array" so as to express 1 as a combination of 26 and 7. Beginning with the next-to-last line in the array (which has the remainder 1), we rearrange it to read

$$1 = 5 - (2 \times 2)$$

This result, which is obviously true, shows 1 written as a combination of 5 and 2. Now, using the previous line with the remainder 2 written in the form $2 = 7 - (1 \times 5)$, we substitute for 2 in the above equation and get

$$1 = 5 - 2 \times [7 - (1 \times 5)] \quad \text{or} \quad 1 = (3 \times 5) - (2 \times 7)$$

This result, which is obviously true, now shows 1 written as a combination of 7 and 5. Finally, the first line of the array of calculations says that $5 = 26 - (3 \times 7)$, so substituting for 5 in the last equation displayed gives

$$1 = (3 \times [(26 - (3 \times 7)]) - (2 \times 7)$$

Tidying up this expression gives

$$1 = 3 \times 26 - 11 \times 7$$

another result which is easy to verify.

By working back up the array in the manner just outlined, we are able to express the g.c.d. of 26 and 7, which is 1, as a combination of 26 and 7. It is the presence of the 26 in $1 = 3 \times 26 - 11 \times 7$ which is all important. It allows us to say that

$$-11 \times 7 = -3 \times 26 + 1$$

or equivalently that

$$(-11 \times 7) \bmod 26 = 1$$

This says that when the number 7 is multiplied by -11 and the result reduced modulo 26, the number 1 is obtained. Thus -11 is a multiplicative inverse of 7 modulo 26. Since $-11 \bmod 26 = 15$, the multiplicative inverse of 7 modulo 26 is 15.

$$15 \times 7 = 105 = 4 \times 26 + 1$$

shows that

$$(15 \times 7) \bmod 26 = 1$$

This method of finding the multiplicative inverse of a number modulo another number may strike you as a form of black magic. Even if it doesn't, you will almost certainly think that it is laborious or tedious because you must first calculate the array in the forward direction and, when this is done, you must work your way back up through it to write the number 1 as a combination of the original two numbers. For a human being all this may indeed be tiresome, but for a properly programmed computer it is not. There is a modern version of Euclid's algorithm, called the extended Euclidean algorithm, which allows both the forward process and the backward process to be done simultaneously—or in "one pass," to use modern jargon. This is fantastic as it gives the g.c.d. and expresses it as a linear combination of the two numbers.

I show (without explanation, sorry!) the relevant array of calculations when this extended algorithm is applied to the numbers 26 and 7, in the hope that you might be curious and look up a textbook to learn what on earth is going on. Here it is:

1	0	26	–
0	1	7	–
1	−3	5	3
−1	4	2	1
3	−11	1	2
–	–	0	2

Mysteriously, the second-to-last row reveals that (26, 7) = 1 and that

$$3 \times 26 + (-11) \times 7 = 1$$

The extended Euclidean algorithm applied to the numbers 676 and 159 yields the array

1	0	676	–
0	1	159	–
1	−4	40	4
−3	13	39	3
4	−17	1	1
–	–	0	39

The second-to-last row reveals that (676, 159) = 1 and that

$$4 \times 676 + (-17) \times 159 = 1$$

Should you require the multiplicative inverse of 159 modulo 676 then this equation leads you to it almost immediately. Rewriting the equation as

$$(-17) \times 159 = -4 \times 676 + 1$$

shows that -17 is a multiplicative inverse of 159 modulo 676. Since $-17 \bmod 676 = 659$, the multiplicative inverse of 159 modulo 676 is 659.

Knowing that 659 is the multiplicative inverse of 159 modulo 676 allows the deciphering rule corresponding to the enciphering rule

$$C = (159P + 580) \bmod 676$$

to be calculated. Adding -580 to both sides of this equation gives

$$(C - 580) = 159P \bmod 676$$

or

$$159P = (C + 96) \bmod 26$$

since $-580 \bmod 676 = 96$. Multiplying both sides of this equation by 659 (which is the multiplicative inverse of 159), and reducing all products modulo 676, gives

$$P = (659C + 396) \bmod 676$$

since $659 \times 96 = 63{,}264 = 93 \times 676 + 396$. Thus

$$P = (659C + 396) \bmod 676$$

is the deciphering rule corresponding to the enciphering rule

$$C = (159P + 580) \bmod 676$$

Appendix D

The Euler ϕ-function and the Euler-Fermat Theorem

Leonhard Euler wrote $\phi(n)$ for the number of positive integers less than a positive integer n which are relatively prime to n. Two natural numbers are *relatively prime* to each other if 1 is the only factor they have in common. For example, $\phi(12) = 4$ since only the four integers 1, 5, 7 and 11 in the set {1, 2, 3, 4, 5, 6, 7, 8, 9, 10, 11} of positive integers less than 12 are relatively prime to 12.

Can you make any observations about the function $\phi(n)$ by examining the following table?

Values of $\phi(n)$ for $n = 2$ to 32

n	$\phi(n)$	n	$\phi(n)$	n	$\phi(n)$	n	$\phi(n)$	n	$\phi(n)$
2	1	8	4	14	6	21	12	27	18
3	2	9	6	15	8	22	10	28	12
4	2	10	4	16	8	23	22	29	28
5	4	11	10	17	16	24	8	30	8
6	2	12	4	18	6	25	20	31	30
7	6	13	12	19	18	26	12	32	16

Here are some values of $\phi(n)$ when n is a prime number:

$$\phi(7) = 6 \ ; \ \phi(13) = 12 \ ; \ \phi(31) = 30$$

Check some more prime values. Hopefully, your observations will lead you to conjecture that

$$\phi(p) = p - 1$$

whenever p is a prime. This is true in general, since the $p - 1$ positive integers *less* than a prime number p can have only the factor 1 in common with p.

This is a very important rule. When combined with two other rules that are less obvious, $\phi(n)$ can be calculated for any natural number n *provided* the complete factorization of the number n is known. Before I tell you about these rules, let me show you something. Observe that

$$4 = \phi(12) = \phi(3) \times \phi(4) = 2 \times 2$$

but that

$$4 = \phi(12) \neq \phi(2) \times \phi(6) = 1 \times 2$$

Furthermore,

$$6 = \phi(18) = \phi(2) \times \phi(9) = 1 \times 6$$

but

$$6 = \phi(18) \neq \phi(3) \times \phi(6) = 2 \times 2$$

It would appear that if $n = r \times s$ then $\phi(n) = \phi(r) \times \phi(s)$ sometimes, but not always. Here is the secret about the property the factors r and s must possess:

$$\phi(n) = \phi(r \times s) = \phi(r) \times \phi(s) \text{ whenever } (r, s) = 1$$

This is a very important rule, but it is not so easy to prove.

The remaining important rule, which is much easier to prove, is that

$$\Phi\ (p^{k}) = p^{k} - p^{k-1}$$

when p is a prime and k is any positive integer. For example, this rule says that

$$\phi(32) = \phi(2^5) = 2^5 - 2^4 = 32 - 16 = 16$$

This is correct, since the sixteen odd numbers 1, 3, 5, . . . , 27, 29, 31 are the only numbers less than 32 which are relatively prime to 32. All the other even numbers share the factor 2 with 2^5.

The three rules described are all that is needed to calculate $\phi(n)$ for any positive integer n once its factorization is known.

Example: To calculate $\phi(9000)$, we first factor 9000 to find that

$$9000 = 2^3 \times 3^2 \times 5^3$$

Then since 2^3, 3^2 and 5^3 are relatively prime to one another, we may write

$$\phi(9000) = \phi(2^3) \times \phi(3^2) \times \phi(5^3)$$

Since

$$\begin{aligned} \phi(2^3) &= \quad 2^3 - 2^2 = 8 - 4 = 4 \\ \phi(3^2) &= \quad 3^2 - 3^1 = 9 - 3 = 6 \\ \phi(5^3) &= \quad 5^3 - 5^2 = 125 - 25 = 100 \end{aligned}$$

it follows that

$$\phi(9000) = 4 \times 6 \times 100 = 2400$$

Perhaps you might like to verify this result directly.

The Euler-Fermat Theorem (1736): If n is any natural number then

$$a^{\phi(n)} \equiv 1 (\text{mod } n)$$

for every natural number a satisfying $(a, n) = 1$.

Equivalently,

$$n | a^{\phi(n)} - 1$$

for every natural number a satisfying $(a, n) = 1$.

Example: If $n = 18$ then $\phi(n) = \phi(18) = 6$. The six numbers less than 18 relatively prime to it are 1, 5, 7, 11, 13 and 17. Letting a be each of these numbers in turn, the Euler-Fermat theorem says that

$$\begin{aligned}
1^6 &\equiv 1 \text{ (mod 18)} \text{ or that } 18|1^6 - 1\\
5^6 &\equiv 1 \text{ (mod 18)} \text{ or that } 18|5^6 - 1\\
7^6 &\equiv 1 \text{ (mod 18)} \text{ or that } 18|7^6 - 1\\
11^6 &\equiv 1 \text{ (mod 18)} \text{ or that } 18|11^6 - 1\\
13^6 &\equiv 1 \text{ (mod 18)} \text{ or that } 18|13^6 - 1\\
17^6 &\equiv 1 \text{ (mod 18)} \text{ or that } 18|17^6 - 1
\end{aligned}$$

Checks:

$$\begin{aligned}
1^6 - 1 &= 0 = 18 \times 0\\
5^6 - 1 &= 15{,}624 = 18 \times 868\\
7^6 - 1 &= 117{,}648 = 18 \times 6{,}536\\
11^6 - 1 &= 1{,}771{,}560 = 18 \times 98{,}420\\
13^6 - 1 &= 4{,}826{,}808 = 18 \times 268{,}156\\
17^6 - 1 &= 24{,}137{,}568 = 18 \times 1{,}340{,}976
\end{aligned}$$

When $n = p$ where p is a prime, $\phi(n) = \phi(p) = p - 1$. In this particular case, the Euler-Fermat theorem says that:

If p is any prime number then

$$a^{p-1} \equiv 1 \pmod{p}$$

for every natural number a satisfying $(a, p) = 1$.

This is Fermat's little theorem (FLT) of 1640. Thus the Euler-Fermat theorem is a generalization of the FLT to all natural numbers. A tremendous achievement!

Acknowledgments

We wish to thank Pat "Herring" Ahern, Tom Barry, Kevin J. Kelly, Tom Laffey, Gerard Murphy and Denise O'Driscoll, who all read drafts of the manuscript. They generously devoted their holiday time to this duty, and we appreciate the many improvements of a mathematical and grammatical nature that they suggested. We are also indebted to Kevin McCarthy for his very careful reading of the first printing of this book, and thank him for confessing that the task reawakened his interest in mathematics.

Our particular thanks must go to Peter Carson of Profile Books, who extended the invitation to write this book. We cursed him privately many times for so doing, but the fates punished him for the sin of making us an offer we couldn't, in conscience, refuse. He must have been forced many times to call on his reserves of patience and tact to encourage two amateurs in their struggle at pretending to be authors. He was forever helpful but never interfering, expending many hours in correspondence and reading with obvious care a number of drafts of different chapters. The fact that this book has seen the light of day owes much to his efforts. *Míle buíochas.*

A thousand thanks also to John Woodruff for his superb copyediting and the numerous improvements suggested by his knowledge of mathematical history. He has been a true champion of the reader. Thanks also to Kate Griffin and Andrew Franklin, both of Profile, for their wonderfully open ways and warm enthusiastic encouragement. A special thanks to Andrew's daughter, Miriam. Knowing that she enjoyed the puzzles and "couldn't wait to see the book" was an inspiration on the many occasions when our confidence was flagging.

We also wish to acknowledge the painstaking "behind-the-scenes" professionalism and efforts of Stephen Brough of Profile and Jonathan Harley, and we thank them both for the splendid way they have produced this book.

We must commend Profile on the courage they have shown by publishing a book that contains as much mathematics as this one does. Only time will tell if they are as mad as we are in thinking that the general reading public is no longer intimidated by exposition which dares to contain technical detail.

We were very heartened that this may indeed be the case when we learned that the American publishing house of Workman, far from being unsettled by the mathematical content of this work, was keen to place the book before the American reader. We thank those responsible for this bold decision, in particular the late Sally Kovalchick, who initiated the project with such enthusiasm and Suzanne Rafer, who saw it through with an equal amount of enthusiasm. We hope that their faith is rewarded by a favorable reception of the book.

For the preparation of the American edition we had the energetic and wholehearted assistance of Richard Simon. It was he who suggested a restructuring of the earlier chapters of the Irish/English edition that would treat the reader to considerably more of the personal story than we presented in the corresponding chapters of the original. We were more than happy to consent to this, realizing that by delaying the appearance of mathematics, we might better motivate our readers to undertake the slow, steady climb through this material once they encounter it. If the result of this new structure is that the reader now attains that summit from where the personal and mathematical stories are clearly seen to blend into one overall tale, then the reworking will not have been in vain.

As a writer of English as it is spoken in America, Richard suggested many changes, in particular in relation to idioms of Irish speech. By removing what were sure to be unfamiliar modes of expression, he has saved us from possible obscurity and helped us achieve greater clarity. Although motivated by the need for transparency, he was careful to ensure that the reader not be denied the realization that Sarah's voice is not that of an American, and that she be heard as she is. For this, and much more, we thank him.

Bibliography

Bell, E.T. *Men of Mathematics*. New York: Simon & Schuster, 1937.

Cipra, Barry. *What's Happening in the Mathematical Sciences*. vol. 3. Providence, RI: American Mathematical Society, 1996.

Hardy, G. H. *A Mathematician's Apology*. Cambridge, Eng.: Cambridge University Press, 1940.

Hoffman, Paul. *The Man Who Loved Only Numbers*. London: Fourth Estate, 1998.

Koestler, Arthur. *The Act of Creation*. London: Picador, 1964.

Roberts, Joe. *Lure of the Integers*. Washington, DC: Mathematical Association of America, 1992.

Singh, Simon. *The Code Book*. London: Fourth Estate, 1999.

———. *Fermat's Last Theorem*. London: Fourth Estate, 1997.

Stoll, Clifford. *The Cuckoo's Egg*. London: Doubleday, 1989.

Young, Robert M. *Excursions in Calculus*. Washington, DC: Mathematical Association of America, 1993.

Index